Gestaltung hybrider Mensch-Maschine-Systeme/Designing Hybrid Societies

Series Editor
Angelika C. Bullinger-Hoffmann, Chemnitz, Germany

Veränderungen in Technologien, Werten, Gesetzgebung und deren Zusammenspiel bestimmen hybride Mensch-Maschine-Systeme, d. h. die quasi selbstorganisierte Interaktion von Mensch und Technologie. In dieser arbeitswissenschaftlich verankerten Schriftenreihe werden zu den Hybrid Societies zahlreiche interdisziplinäre Aspekte adressiert, Designvorschläge basierend auf theoretischen und empirischen Erkenntnissen präsentiert und verwandte Konzepte diskutiert.

Changes in technology, values, regulation and their interplay drive hybrid societies, i.e., the quasi self-organized interaction between humans and technologies. This series grounded in human factors addresses many interdisciplinary aspects, presents socio-technical design suggestions based on theoretical and empirical findings and discusses related concepts.

More information about this series at http://www.springer.com/series/16273

Elisabeth Schmidt

Effects of Thermal Stimulation during Passive Driver Fatigue

With a foreword by
Prof. Dr. Angelika C. Bullinger-Hoffmann

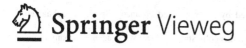 Springer Vieweg

Elisabeth Schmidt
Munich, Germany

Dissertation Chemnitz University of Technology, Germany, 2019

ISSN 2661-8230   ISSN 2661-8249   (electronic)
Gestaltung hybrider Mensch-Maschine-Systeme/Designing Hybrid Societies
ISBN 978-3-658-28157-1   ISBN 978-3-658-28158-8   (eBook)
https://doi.org/10.1007/978-3-658-28158-8

This Springer Vieweg imprint is published by the registered company Springer Fachmedien Wiesbaden GmbH part of Springer Nature.
The registered company address is: Abraham-Lincoln-Str. 46, 65189 Wiesbaden, Germany

Geleitwort

Die zunehmende Automatisierung des Fahrens hat zum Ziel, den Fahrer merklich von der Fahraufgabe zu entlasten, führt jedoch auch zu Ermüdungseffekten. Elisabeth Schmidt widmet ihre Dissertation in diesem Kontext der Beeinflussung des Fahrerzustands durch gezielte Kühlung verschiedener Körperregionen und der Messung des Fahrerzustands.

Mit ihrer Arbeit zu den Effekten von Kühlung auf die passive Müdigkeit von Fahrern führt sie mit drei Fahrsimulatorstudien eine sehr differenzierte Betrachtung des Potenzials der Kühlung des Fahrers durch Luftströmung durch. Ihre Ergebnisse geben klare Hinweise darauf, dass eine zustandsorientierte Klimatisierung als vierminütige Kühlung des Oberkörpers gestaltet werden sollte. Sie zeigen aber auch, dass deren Potenzial bei wiederholter Anwendung nicht überbewertet werden darf. Bezüglich der Akzeptanz einer automatisierten Kühlung findet Frau Schmidt heraus, dass diese hoch ist, so die Fahrer über den Zweck der Kühlung informiert sind.

Die Studien von Frau Schmidt sind im Fahrsimulator durchgeführt, bieten aber dennoch aufgrund ihres differenzierten Vorgehens eine gute Grundlage für die Entwicklung, Gestaltung und Evaluation einer automatisierten Kühlung in der Praxis.

Während die Dissertation auf passive Müdigkeit zugeschnitten ist, bieten die erarbeiteten Ergebnisse und die transparent diskutierten offenen Fragestellungen Potenzial für die Erforschung anderer Müdigkeitsformen.

Ich wünsche Elisabeth Schmidt daher zahlreiche interessierte Leserinnen und Leser aus Wirtschaft und Wissenschaft – und zahlreiche Fahrerinnen und Fahrer zunehmend automatisierter Fahrzeuge, die aufgrund automatisierter Kühlung wach und sicher ihr Ziel erreichen!

Angelika C. Bullinger-Hoffmann

Chemnitz, im Mai 2019

Acknowledgements

My greatest thanks goes to my thesis advisor Prof. Angelika C. Bullinger-Hoffmann. Your thoughtful guidance has been invaluable throughout my research. I am grateful that you supported me in making the most out of my collaboration with the Technische Universität Chemnitz.

I also thank Prof. Klaus Bengler for agreeing to be the second corrector of this thesis and your feedback on the documentation.

Thanks also goes to my supervisor Ralf Decke for your support and help with all administrative hurdles. Thanks to Stefan Wiedemann and your team for your guidance and technical input related to all aspects of automotive air conditioning and your feedback. I also thank Dr. Alexander König for introducing me to the techniques of driver state measurement using physiological signals.

Thank you, Sigrid, for your feedback and advice throughout the thesis. Moreover, your research results of sensory stimulation have created an important basis for this endeavor.

I would like to express my gratitude to Ronee, Hanna, Oliver, Sonja, Florian, Svenja, Michael und Flo from the MMI team for integrating me into your team and on top of that, providing me with the best work desk in your office!

A great thanks goes to Andrea Schöps and Michael Hauser who supported the studies during many long days in the driving simulators.

To André, Dorothea, Patrick, Max, Daniel, Katharina and Svenja, our conversations on design of experiments, the sharing of your research approaches and viewpoints have been inspiring for me. Thank you for recommending literature, lending your books and for being such respectful and kind office-mates during my time in Chemnitz.

The conduction of my research would not have been possible without the participation of hundreds of volunteers who spent hours of driving on probably the most monotonous highway simulation in the world. Thank you for your help.

I would like to thank Nina Kauffmann, Judith Ochs, Nina Brouwer, Michael Festner, Dr. Julia Niemann and Dr. Sebastian Hergeth for the frequent discussions on study designs and the hurdles associated with them.

Thank you, Aljoscha, Stephanie, Markus and Carolyn and many more, for distracting me from research. It is invaluable to have such great friends that I can always count on and who give honest advice.

A special thanks goes to my sister Franziska. I always appreciate your opinion on diverse matters, your ideas and invitations.

To Jacob, you deserve a great deal of gratitude for the hours of proofreading spent on all publications and eventually this thesis. Thank you for being my adventure companion and supporting me in everything I do.

Finally, I would like to thank my caring parents Anna and Anton, who deserve all of the credit for encouraging my educational endeavors and providing me with everything I need to realize them.

Elisabeth Schmidt

Overview of Contents

Table of Contents

List of Abbreviations

Abbreviation	Explanation
A	answer
AC	air conditioner
ADACL	Activation Deactivation Adjective Checklist
ANS	autonomic nervous system
ASHRAE	American Society of Heating, Refrigerating and Air-Conditioning Engineers
BF	breathing frequency
bpm	beats per minute
C	Celsius
CID	central information display
cm	centimeter
CO_2	carbon dioxide
COMB	combination of light, sound, scent and climate
CONT	control condition without cooling
COOL	condition with cooling
DWA	driving without awareness
ECG	electrocardiogram
EEG	electroencephalogram
e.g.	exempli gratia

EOG	electrooculogram
HF	high frequency component of HRV
HMI	human-machine interaction
HR	heart rate
HRV	heart rate variability
Hz	Hertz
i.e.	id est
km/h	kilometers per hour
KSS	Karolinska Sleepiness Scale
LDA	linear discriminant analysis
LF	low frequency component of HRV
m	meter
M	mean
Mdn	median
min	minute
mm	millimeter
ms	millisecond
n	sample size
PERCLOS	percentage of eye closure
PNS	parasympathetic nervous system
ppm	part per million
rel	relative
RMSSD	root mean square of successive differences
ROC	receiver operating characteristic

RQ	research question
s	second
S	Siemens
SAE	Society of Automotive Engineers
SCL	skin conductance level
SD	standard deviation
SDNN	standard deviation of normal-to-normal intervals
SE	standard error
Se-6	microsiemens
SNS	sympathetic nervous system
SR	sleep-related
SSS	Stanford Sleepiness Scale
SVM	support vector machine
t	time
TR	task-related
UEQ	User Experience Questionnaire
U.S.	United States
UX	user experience
vs.	versus
y	years

List of Figures

List of Tables

Abstract

Driver fatigue accounts for a large portion of vehicular crashes. Aside from an increased safety risk, fatigue also raises discomfort while driving. There has been initial research on countermeasures against driver fatigue and several fatigue models highlight the importance of the causal factors of fatigue when conceptualizing countermeasures against fatigue.

Thermal stimulation has been suggested as a countermeasure against passive fatigue caused by task underload. There exist few driving studies, however, that investigate the use of thermal stimuli as fatigue countermeasures. These studies used different stimulus settings, such as stimulus temperature, duration and because cooled body parts. In addition the degree of sleep deprivation in their samples varied. Thus, these studies yield inconclusive results. From the existing insights, it remains unclear under which thermal conditions fatigue can be mitigated. Furthermore, most of the studies did not address the issue of reduced thermal comfort when applying thermal stimulation. Therefore, the main research focus of this thesis is directed to the effect of thermal simulation in mitigating passive fatigue of non-sleep-deprived drivers while maintaining the drivers' comfort.

This dissertation investigated the effect of thermal stimulation in a series of driving simulator experiments. To induce passive fatigue in the drivers, a monotonous highway drive with very little traffic was chosen. This traffic scenario was tested in a simulator study (n=46) to demonstrate that it could induce sufficiently high subjective fatigue. An increase in heart rate variability and a decrease in breathing frequency were measured during the monotonous traffic scenario which confirms onset of fatigue.

Once passive fatigue was successfully induced, the effects of a 6-minute thermal stimulus of 17 °C, directed to the upper body of the drivers were analyzed (n=34). Responses to the stimulus from the subjects included significantly decreased subjective fatigue ratings, an increase in pupil diameter and an increase in skin conductance which are indicators of a short-term activation of the sympathetic nervous system.

In follow-up studies, the effects of different stimulus durations (2 and 4 minutes, n=33) and lower leg cooling (n=42) on passive driver fatigue were investigated. The results show that 4-minute cooling yielded more pronounced effects in heart rate and pupil diameter variation than the 2-minute cooling. The 4-minute lower leg cooling did not affect any of the recorded physiological parameters, however, the drivers felt subjectively more awake.

These studies showed that sufficiently intensive cooling of the upper body of the drivers reduced passive fatigue in the short-term.

The scope of the research included the development of a fatigue-based triggering of climatic changes in the vehicle cabin. In order to develop a regression model for the detection of fatigue, a secondary data analysis of three driving simulator studies with monotonous highway drives was conducted. With a logistic regression of the input parameters heart rate variability, skin conductance level and pupil diameter, fatigue could be detected with an accuracy of 77 %.

The last simulator study employed the regression model to trigger the thermal stimuli based on the drivers' fatigue level (n=94). The study revealed that a 4-minute thermal stimulus of 15 °C directed towards the drivers' upper body, only affected physiological fatigue during the first and second applications. The third application of the stimulus did not yield a significant activation.

A further result of this thesis was the importance of a transparent interaction strategy for the fatigue countermeasure, which includes rationale for the suggested cooling. The strategy should also leave the autonomy of the drivers to accept or reject the suggested countermeasure.

Concluding, the thesis contributes the following insights to the research field of driver fatigue. The studies of the thesis showed that thermal stimulation of the upper body caused physiological and subjective effects which can be associated with a short-term sympathetic activation. In contrast to upper body cooling, leg cooling did not yield physiological activation. Furthermore, the study on repeated stimuli showed that thermal stimulation is subject to habituation.

Keywords: fatigue, fatigue countermeasures, thermal stimulation, vitalization

1 Introduction

Driver fatigue is a major cause of road accidents. As naturalistic driving data of more than 3500 participants from the U.S. show, fatigue increased the risk of a crash by 3.4 times compared to alert driving (Dingus et al., 2016). In an earlier analysis of naturalistic driving data, it was found that 20 % of single-vehicle crashes and 23 % of near-crashes are related to fatigue (Dingus et al., 2006).

Several researches have proposed definitions for the term fatigue. Schmidtke (1965, p.18) summarized these definitions and concluded that

> "[...] fatigue is a consequence of earlier strain that causes reversible performance reductions." (translated from German by the author)

It furthermore reduces motivation, increases perceived exertion and disrupts personality functions (Schmidtke, 1965). Symptoms of fatigue in general can be coordination, concentration, attention and perceptual disorders and even disorders of social relationships (Schmidtke, 1965). In the driving context, Brown (1994) concluded that more frequent and longer eyelid closures caused by fatigue and the subconscious redirection of attentional resources to inner thought processes is the reason for increased impairment of the driver. Not only does fatigue lead to impairment, it also causes discomfort (Van Veen, 2016; Zhang et al.,1996).

While the feeling of fatigue has a warning and protective function, similar to the feeling of pain, the feeling of fatigue and the actual immanent fatigue are not identical. There are examples of highly straining conditions in which a feeling of fatigue is absent (Schmidtke, 1965).

The problem of driver fatigue has been recognized by both academia and industry. Researchers have proposed methods to measure fatigue based on driving data or physiological signals (Friedrichs and Yang, 2010; Patel et al., 2011; Sandberg et al., 2011). Automobile manufacturers offer in-car

© Springer Fachmedien Wiesbaden GmbH, part of Springer Nature 2020
E. Schmidt, *Effects of Thermal Stimulation during Passive Driver Fatigue*,
Gestaltung hybrider Mensch-Maschine-Systeme/Designing Hybrid Societies,
https://doi.org/10.1007/978-3-658-28158-8_1

fatigue detection systems which are based on steering behavior of the driver (Volkswagen AG, 2018), lane keeping behavior (Volvo Cars, 2018) or gaze behavior (Toyota, 2018). A detailed review of state-of-the-art in-car systems or retrofit-solutions offered by the industry is provided by Schmidt (2018). Schmidt (2018) also summarized that most of the offered systems lack scientific validation of the detection accuracy. A review of a selection of current fatigue detection devices performed by Dawson et al. (2014, p. 1) revealed that "none of the current technologies met all the proposed regulatory criteria for a legally and scientifically defensible device."

In addition to the research efforts on fatigue detection, there has also been scientific and industrial work towards measures of fatigue mitigation. Van Veen et al. (2014) provided a review of studies suggesting different countermeasures for different types of fatigue. Among these, thermal stimulation is listed as a countermeasure against fatigue arising from monotony. An analysis of driving studies investigating thermal stimulation, though, yielded that the effectiveness of thermal stimulation depends strongly on the stimulus settings, such as stimulus temperature, duration and cooled body parts (Schmidt and Bullinger, 2019).

The idea of influencing driver states has lately been picked up by automobile manufacturers. Audi AG (2018), BMW AG (2019) and Daimler AG (2018) have introduced concepts to mitigate critical drivers' states by means of adjusting the vehicle's interior settings, such as climate and light conditions. These systems also show limited scientific evidence for their effectiveness, as with the fatigue detection systems mentioned above. On one hand, this is because the methods used in the automobile industry are not effective against fatigue per se, but rather against one type of fatigue arising under certain conditions (Van Veen et al., 2014), which are rarely considered in fatigue detection systems. On the other hand, even if applied under the appropriate conditions, the subjective and objective effect of thermal stimulation varies for different climate settings (Schmidt and Bullinger, 2019).

From the few existing driving studies investigating thermal stimulation (Landström et al., 1999, 2002; Reyner and Horne, 1998; Schwarz et al., 2012; Van Veen, 2016; Wyon et al., 1996) the results on fatigue mitigation remain inconclusive.

1.1 Motivation and Relevance

This thesis focuses on questions relating to **the effect of thermal stimulation as a possible countermeasure** against passive driver fatigue, arising from task underload. As stated above, fatigue is not only a risk for traffic safety, it also increases discomfort while driving. This has been shown in the classification analysis of discomfort of Zhang et al. (1996) as well as the fatiguing process of Van Veen (2016). Van Veen (2016, p. 8) describes the fatiguing process using four stages 1) imperceptible stage, 2) discomfort, 3) impairment and 4) sleep.

This thesis **aims to address the discomfort-related aspect (stage 2) of driver fatigue.** Kuorinka (1983, p. 1090) defined that

> "[...] discomfort is a phenomenon of perception. Related phenomena are fatigue, perceived exertion and pain. [...] discomfort refers to the unpleasant, often painful, physical sensation."

Bubb et al. (2015, p. 149) concluded that discomfort relates to the pragmatic quality of a product. The pragmatic quality is a dimension of user experience (UX) according to the model of Hassenzahl (2007). Therefore, the reduction of discomfort arising from fatigue while driving would increase the pragmatic quality of a product and hence improve the user's experience.

Comfort has been defined as the

> "absence of discomfort, [...] a state of no awareness at all of a feeling" (Hertzberg, 1972, p. 41) and, in addition, the "luxury appraisal of features" (Bubb et al., 2015, p. 146).

Bubb et al. (2015, p. 149) further explained that comfort relates to the hedonic quality of a product.

As stated above, the discomfort-related aspect of driver fatigue and its mitigation by means of thermal stimulation is central in this thesis. As it is easy to imagine, the thermal environment is directly linked to thermal comfort (Parsons, 2002). This could potentially cause **a trade-off in the design of suitable thermal stimulation** to both reduce fatigue and maintain the driver's thermal comfort. Little is gained from a system that reduces fatigue using extreme sensory stimulation, but causes direct discomfort. While existing studies on thermal stimulation as a countermeasure against fatigue have measured subjective and objective indicators of fatigue (Landström et al., 1999, 2002; Reyner and Horne, 1998; Schwarz et al., 2012; Wyon et al., 1996), they neglected the (dis-)comfort-related aspects of thermal stimulation. Only Van Veen (2016) has explicitly recorded drivers' comfort perceptions after short-term cooling of the hands has been applied. Therefore, this thesis aims to address the discomfort-related aspect of thermal stimulation along with a thorough investigation of the effectiveness in mitigating passive fatigue.

As stated above, the current insights on thermal stimulation are inconclusive because previous studies have applied different stimulus temperatures, durations and cooled body parts to differently characterized samples. Therefore, it is not possible to draw general conclusions on the conditions under which thermal stimulation is effective against passive fatigue and how long the effect lasts. The body of literature still lacks a thorough investigation of underlying mechanisms, effect strength and effect duration of the stimulus. Insights are furthermore needed in terms of ideal conditions of the stimulus to maintain comfort and in terms of effect behavior upon stimulus repetition.

This thesis contributes to the research field of driver fatigue because it creates knowledge on the **effectiveness of specific countermeasures against fatigue**. Even though, academia and industry still require a scientifically valid means to detect fatigue including its causal factors as stated above, the gained insights on countermeasures are relevant. In recent years, several advances in technologies have the potential to support the accurate detection of fatigue. The first is the rapid progress in sensor technologies that allows, e.g., for measuring peoples movements, behavior, physiological data and sleep cycles with devices such as smartphones or wristbands (Swan, 2012). Second, increasing progress in data analytics and artificial intelligence helps to accurately interpret sensor data (Müller and Bostrom, 2016). Finally, an increasing degree of connectivity of different devices in our environment helps to share, use and act upon the transmitted information (Mazhelis et al., 2012).

Summarizing, the required technological advancements to enable valid fatigue detection are underway, and hence the gained insights on fatigue countermeasures can be integrated.

1.2 Research Questions and Method

The focus of this thesis is to empirically investigate the effect of thermal stimulation as a countermeasure against passive fatigue while driving. In order to reach this goal, human subject studies are required in which thermal stimuli are applied on passively fatigued drivers. To ensure study participants are sufficiently fatigued, a pre-test is required which answers the first research question (RQ):

> RQ 1: Which degree of passive fatigue can be induced by means of task underload in drivers according to subjective and objective measures?

To answer this question, a pre-test in a driving simulator is performed to investigate the fatigue inducing effect of monotonous traffic scenarios on subjective and objective fatigue measures. The expected results are recommendations for the suitability of certain traffic simulations and durations for the induction of passive fatigue. The study results also provide an estimate on how fatigued the participants become on average.

The main part of this thesis consists of studies directed to answering the second RQ:

> RQ 2: Which effects do short-term thermal stimuli have on the drivers' passive fatigue?

The main studies of this thesis, which address RQ 2, consist of three driving simulator studies, each investigating a different setting of the vehicle's air conditioner (AC). Fatigue is induced in the participants by means of the driving simulation on a monotonous road previously described. To explore each setting of the AC, studies are conducted employing within- or between-subject designs. These studies compare both subjective and physiological fatigue measures to a thermo-neutral control condition or control group, respectively. In doing so, the results can be compared in terms of effectiveness of different settings and recommendations for suitable stimuli characteristics can be determined. Attention is paid to addressing the trade-off between the effect of the stimulus on fatigue and its effect on thermal comfort.

The objective measurement of fatigue is an important part for the practical application of thermal stimulation for two reasons. First, it is essential to determine the exact point in time when thermal stimulation should be applied. Second, fatigue measurement can be used for continuously evaluating the effectiveness of the countermeasure in terms of fatigue reduction without requiring the drivers to respond to questionnaires. Therefore, the third RQ follows:

> RQ 3: At which accuracy can passive fatigue be detected based on physiological measures?

In order to investigate how accurately physiological indicators can predict fatigue, an analysis is performed on a series of simulator studies that provide the training data. This training data consists of subjective fatigue ratings by the study participants along with their physiological measurements at the time of the subjective assessment. Several input variable combinations and prediction methods are tested against each other in this step. As a result of this analysis, a regression model for fatigue detection is developed along with its classification accuracy.

Having created a knowledge basis of the above three RQs, it is possible to implement the findings in a final study. This study incorporates the online fatigue measurements based on physiological data and the fatigue countermeasure, which is triggered when fatigue is detected. This study aims to provide insights in order to answer the fourth RQ:

> RQ 4: Which effect do 4-minute thermal stimuli at 15 °C have on drivers' fatigue, when applied in response to the drivers' fatigue level?

To answer RQ 4, the fatigue detection and the recommended thermal setting are tested in a closed-loop. In this test, the cold stimulus is repeatedly triggered each time fatigue is detected. This allows for the observation of repeated stimuli and the analysis of their effects, while avoiding the subjective verbal assessment of fatigue which facilitates the preservation of the monotony of the drive.

To complete the research on a fatigue management system, which detects and counteracts fatigue, a final question is directed towards a future human-machine interaction (HMI) with such a system:

> RQ 5: Which effect do different interaction strategies have on automation trust and acceptance of the fatigue management system?

To investigate this question the above mentioned closed-loop study (RQ 4) employed a between-subject design, to incorporate different interaction strategies for the fatigue management system for each experimental group. After driving with the fatigue management system, study participants are asked to subjectively rate trust in, and acceptance of, the system. The expected study results allow for suggestions for suitable HMI concepts for fatigue management systems.

1.3 Structure of Thesis

The parts of this thesis are structured as shown in Figure 1. Chapter 1 begins with the **motivation and relevance** of this research (Chapter 1.1). Subsequently, the **research questions and methods** are outlined (Chapter 1.2), followed by this **overview of the thesis structure** (Chapter 1.3).

Chapter 2 provides **a review of related work in the field of driver fatigue** (Chapter 2.1) **and effects of thermal environments on passive fatigue** (Chapter 2.2) and a **conclusion from the review** (Chapter 2.3).

The first part (Chapter 2.1) summarizes the literature on driver fatigue, i.e., different causes and types of driver fatigue as well as suggested countermeasures for different fatigue types. It also includes a brief review of the reported methods to subjectively measure driver fatigue and the physiological indicators of fatigue. This part closes with an overview of existing algorithms to detect and quantify fatigue based on objective indicators.

The second part (Chapter 2.2) is dedicated to the theoretical and empirical foundations of thermal stimulation. This includes the underlying neurological processes in thermal perception and details why thermal stimulation may be suitable as a fatigue countermeasure.

The conclusions from the review of related work are drawn in Chapter 2.3.

The **tools used and methods** for the subsequent studies are described in Chapter 3, including overall study design, driving simulator and apparatus specifications, provided stimuli, recruitment of participants as well as the procedure for data recording and analysis.

Chapter 4 discusses the **fatigue induction in driving simulator studies** by means of monotonous traffic scenarios. The chapter details the design, results and discussion of a simulator study, which explores the feasibility to induce passive fatigue. With data from the same study, correlation analyses between subjective and objective fatigue indicators are performed that serve as an initial screening for promising fatigue predictors.

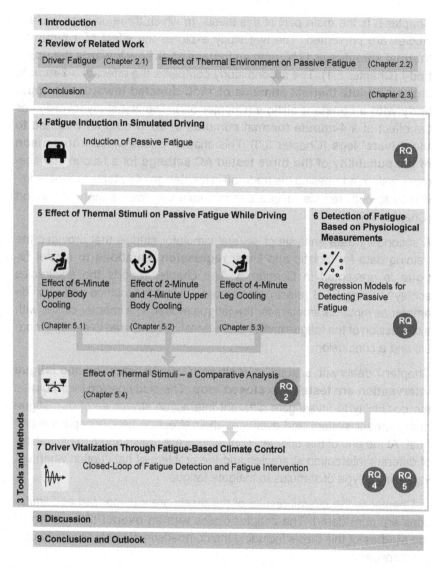

Figure 1: Structure of thesis
Source: *Own illustration*

Chapter 5 is the main part of this thesis, in which three driving simulator studies are presented. The first study explores the effect of a **6-minute thermal stimulus at 17 °C with an increased airflow towards the upper body** (Chapter 5.1). The second study compares the effect of a **2-minute and a 4-minute thermal stimulus of 15 °C directed towards the upper body with a constant airflow** (Chapter 5.2). The last study investigates the effect of a **4-minute thermal stimulus of 15 °C cold air directed to the drivers' legs** (Chapter 5.3). This chapter closes with a **comparison of the suitability of the three tested AC settings** for a fatigue management system addressing the trade-off in the design of suitable thermal stimulation to both reduce fatigue and maintain the driver's thermal comfort (Chapter 5.4).

A secondary data analysis of several simulator studies that provided the training data for logistic and linear **regression equations to model fatigue** is presented in Chapter 6. The chapter reports the accuracies achieved using ECG (electrocardiogram), skin conductance and pupil diameter as input parameters for the fatigue model. The chapter closes with a discussion of the fatigue model characteristics compared to existing models and a conclusion.

Chapter 7 deals with a study, in which both **fatigue detection and fatigue intervention are tested in a closed-loop**. The study design incorporates the possibility to investigate different interaction strategies of the fatigue management system and evaluates their effect on user acceptance and trust. At the end of this chapter, conclusions are drawn about the suitability of different interaction strategies and the problem of habituation when using a single type of stimulus to mitigate fatigue.

In Chapter 8, the overall results of the thesis are summarized and limitations are considered. The chapter closes with an **overall discussion of the studies** of this thesis including the comparison of study results to previous research.

Chapter 9 outlines the **practical implications** of the insights from this thesis for the automobile industry. Finally, **directions for future research** are recommended.

2 Review of Related Work

This thesis brings together two different domains of automotive ergonomics: driver fatigue and thermal environment. The next chapters introduce relevant concepts in these domains and review the related work. Chapter 2.1 provides an overview of different types of driver fatigue and methods to quantify fatigue and Chapter 2.2 addresses thermal environments and thermal perception. At the end of Chapter 2.2, the studies relevant to the two domains, driver fatigue and thermal environment, are reviewed. The section closes with a conclusion from the presented related work (Chapter 2.3).

2.1 Driver Fatigue

While Chapter 1 introduced the term fatigue in a general context using the definition of Schmidtke (1965), there exists more specific research detailing the definition and causal factors of different types of fatigue in the context of driving which will be presented in the following. An even more detailed review of causal factors and countermeasures of passive driver fatigue is required because of its central role in this research. A review of measures to counteract and detect passive driver fatigue is important for the understanding of the research gap that is addressed by this thesis.

Chapter 2.1.1 contains a review of different types of driver fatigue as defined by several models from literature. A summary of studies investigating causes and countermeasures of passive driver fatigue is presented in Chapter 2.1.2. In Chapter 2.1.3 and 2.1.4, different subjective and objective fatigue measurement methods as well as fatigue detection models, described in literature, are reviewed.

© Springer Fachmedien Wiesbaden GmbH, part of Springer Nature 2020
E. Schmidt, *Effects of Thermal Stimulation during Passive Driver Fatigue*,
Gestaltung hybrider Mensch-Maschine-Systeme/Designing Hybrid Societies,
https://doi.org/10.1007/978-3-658-28158-8_2

2.1.1 Types of Fatigue

As the reviews of Lal and Craig (2001), May and Baldwin (2009) and Van Veen et al. (2014) postulate, there are different types of driver fatigue. These reviews differentiate between sleep-related (SR) fatigue, task-related (TR) fatigue and physical fatigue. This thesis adds on to the above defined types of fatigue by including the concept of driving without awareness (DWA), and hence expanding the existing driver fatigue models of May and Baldwin (2009) and Van Veen et al. (2014). An overview of the expanded driver fatigue model is shown in Figure 2. Figure 2 also includes a summary of possible countermeasures against each type of fatigue. The countermeasure have, in parts, been adopted from the review of Van Veen et al. (2014).

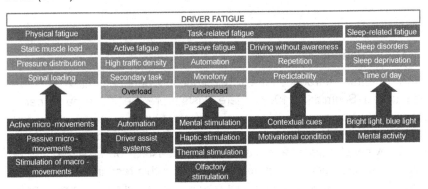

Figure 2:   Driver fatigue model

Source: *Adopted from May and Baldwin (2009) and Van Veen et al. (2014)*

Physical fatigue in driving situations is caused by the static muscle load in prolonged sitting (El Falou et al., 2003; Lal and Craig, 2001; Van Veen et al., 2014). Active or passive micro-movements are supposed to mitigate this type of fatigue (Figure 2). The study of Van Veen et al. (2015) showed for example that participants felt significantly more energetic when the car seat imposed continuous posture variations.

TR fatigue is another type of fatigue that impairs drivers. This type has been suggested by Tejero Gimeno et al. (2006), May and Baldwin, (2009) and Neubauer et al. (2012) to be further distinguishable in fatigue through task overload, known as **active fatigue**, and task underload, known as **passive fatigue**. The literature review of this thesis on fatigue has analyzed that **DWA** should be included in the fatigue model under TR fatigue as this phenomenon shows similar symptoms as fatigue, such as changes in blink behavior (Briest et al., 2006; Karrer et al., 2005).

Task overload can happen, for example, through engagement in secondary tasks and/or in traffic scenarios that require the capturing and processing of a large amount of information and reactions, like when driving in unknown city intersections with high traffic densities (May and Baldwin, 2009). In accordance with the resource theory (Helton and Russell, 2012), such task overload results in a vigilance decrement because the information processing resources are limited. Countermeasures against active fatigue may be automation (Figure 2), given that the operation mode is fully understood by the drivers (Van Veen et al., 2014).

Task underload is characterized by requiring little information processing and little or no actions from the driver. Neubauer et al. (2012, p. 370) explained:

> "Passive fatigue is typically triggered by low workload and monotonous driving tasks."

This can occur when the driving task is automated or predictable, for example, on monotonous highways with little or no traffic (Tejero Gimeno et al., 2006). To decrease fatigue due to monotony, several research efforts have been undertaken. Tejero Gimeno et al. (2006), Desmond and Matthews (1997) and Van Veen et al. (2014) listed secondary tasks and sensory stimulation as possible countermeasures (Schmidt and Bullinger, 2019). A detailed summary of causal factors and countermeasures against passive fatigue is provided in Chapter 2.1.2.

The repetitive and predictable nature of monotonous driving turns the task into an automatic process. This can cause DWA in which drivers do not pay conscious attention to their surroundings and have no recall of the past driving minutes (Brown, 1994; Charlton and Starkey, 2011). Briest et al. (2006) and Karrer et al. (2005) found that passive fatigue and DWA are closely related and show similar symptoms, like changes in blink behavior. When DWA or subconsciously redirecting attentional resources to inner thought processes, inattention due to fatigue can occur without a subjective feeling of fatigue (Brown, 1994). This is also supported by May (2011) who introduced subjective fatigue as a mere symptom of fatigue. Charlton and Starkey (2011, 2013) have suggested contextual cues, for example speed cues, and motivational conditions as countermeasures to be implemented by road safety practitioners as a measure against DWA.

SR fatigue is caused by sleep disorders, sleep deprivation and circadian effects. Some researchers have suggested countermeasures against SR fatigue. Previous research provided evidence that bright light and blue light reduced fatigue and increased reaction times (Lowden et al., 2004; Phipps-Nelson et al., 2009). A different approach was tested by Verwey and Zaidel (1999) who found in a simulator study that subjective sleepiness was reduced and driving performance was enhanced by engaging sleep-deprived drivers in mental activity games.

In practice, driver fatigue often cannot be assigned to just one type of fatigue, because it results from a mix of contextual factors (Lal and Craig, 2001).

2.1.2 Passive Fatigue – Causes and Countermeasures

The revolution of automation (="execution by a machine [...] of a function that was previously carried out by a human" (Parasuraman and Riley, 1997, p. 231)) greatly impacts workplaces in the domains of manufacturing, aviation and automobiles to name just a few (Parasuraman and Riley, 1997). Young and Stanton (2002) describe that automation reduces operator workload and warn about the negative effects of mental underload on performance. In the context of driving, Wang et al. (2017, p. 741) explain that "automated driving systems may reduce active fatigue, but in turn result in passive fatigue".

Because passive fatigue gains more importance in the context of ongoing automation and is relevant for operators and workplaces in various domains, this thesis focuses on this type of fatigue.

Causes

As stated in Chapter 2.1.1 the cause of passive driver fatigue is monotony, which occurs when the driving situation requires low workload (Neubauer et al., 2012). McBain (1970) defined a situation as monotonous when there is no change in stimuli or when the change is predictable. The lack of stimuli in these conditions causes deactivation of the driver allowing fatigue to develop.

By reviewing the literature on causal factors on passive fatigue, three types of factors emerged: factors related to the environment, vehicle and person. Figure 3 schematizes a model for passive driver fatigue which clusters different causes and countermeasures that have been analyzed in previous research.

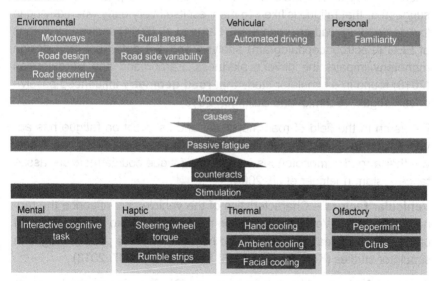

Figure 3:   Passive driver fatigue model
Source: *Own illustration*

Environmental factors – The road scenery is an important factor when researching passive fatigue. As Horne and Reyner (1995) concluded from their surveys, fatigue-related accidents account for a larger proportion of accidents (23 %) on **motorways or other monotonous roads** compared to accidents in general (16 %). The survey of Fell and Black (1997) on driver fatigue incidents in cities compared to **rural areas**, included subjective ratings of tiredness at the start of the trip. Their results showed that 45 % of the drivers did not feel tired before the trip in rural areas whereas for drivers in cities this percentage decreased to 27 %. This means that in the case of a fatigue incident in a rural area, a larger part of the drivers developed inattention due to monotony rather than due to high fatigue levels at the beginning of the drive.

In addition to the surveys of Horne and Reyner (1995) and Fell and Black (1997), the simulator studies of Larue et al. (2011) and Thiffault and Bergeron (2003) confirmed that **road design** and **road side variability** influenced passive fatigue development. Thiffault and Bergeron (2003, p. 390) suggested "that fatigue is likely to manifest itself very early when driving in low demanding road environments". Larue et al. (2011) showed by means of EEG (electroencephalogram) analysis and eye data, that road design monotony impairs the driver's alertness. Farahmand and Boroujerdian (2018) found in a simulator study that **road geometry** influenced passive driver fatigue. (Schmidt and Bullinger, 2019)

Research in the field of road monotony and its effect on fatigue has advanced to the level that algorithms have been developed that can detect whether a road is monotonous as part of a fatigue countermeasure assistance system (Fletcher et al., 2005).

Vehicular factors – Characteristics of the vehicle can influence the monotony of a drive. Vehicle **automation** has been shown to reduce workload on the driver, increase monotony and hence to increase passive fatigue in simulator studies (Körber et al., 2015; Saxby et al., 2007, 2013).

Personal factors – Charlton and Starkey (2013) reviewed literature that suggests that crashes are most likely to happen near the homes of the drivers. They conducted a simulator study, in which participants were required to drive on the same road regularly over a period of three months. They found that **familiarity** resulted in lower ratings of mental demand and inattention blindness.

Countermeasures

Secondary tasks and sensory stimulation have been used to reduce passive fatigue (Van Veen et al., 2014). Neubauer et al. (2012) have proposed interactive technologies as countermeasures. The model in Figure 3 suggests that these tasks can be abstracted to be a type of stimulation as well, namely mental stimulation. Other types of stimulation included in Figure 3 are haptic, thermal and olfactory stimulation.

Mental – Mental activity has been shown in several studies to mitigate passive fatigue. Gershon et al. (2009a) found that an **interactive-cognitive-task** (Trivia) in which drivers were asked questions via speakers in the fields of movies, sports, current events, general knowledge and cuisine while driving manually in a simulator increased physiological arousal and alertness. Similarly, Schömig et al. (2015) and Jarosch et al. (2017) reported lower levels of subjective fatigue and lower PERCLOS (percentage of eye closure) while engaging in a quiz task displayed on a tablet during an automated drive in motion-based simulators. In all three studies, the drivers were required to answer the questions by manually clicking or touching a button.

Haptic – The literature review yielded evidence that haptic stimulation is also suited to mitigate passive fatigue. Wang et al. (2017) tested a system that continuously exerts **torque on a steering wheel** on passively fatigued simulator drivers. They measured physiological arousal and improved driving performance when the haptic system was active. Another possible indication for the fatigue-mitigating effect of haptic stimulation is the result of Merat and Jamson (2013) who investigated the effect of **rumble strips** on long monotonous stretches of straight roads.

Thermal – Van Veen et al. (2014) derive from their literature review that passive fatigue can be reduced by means of thermal stimulation. This was confirmed in a simulator study in which repeated short-term **cooling of the**

driver's hands caused physiological arousal as measured by an increase of heart rate (HR) (Van Veen, 2016). Outside of the driving context, Tham and Willem (2010) found that also **ambient cooling** increases alertness in office occupants and Collins et al. (1996) reported physiological arousal induced by **facial cooling** in a clinical setting. An extensive elaboration on the theory and empirical studies on thermal stimuli is provided in Chapter 2.2.

Olfactory – Previous research suggests that **peppermint** and **citrus** oil can be used as a measure to reduce fatigue. Norrish and Dwyer (2005) and Raudenbush et al. (2001) found that peppermint was effective against fatigue outside of the driving context. Mahachandra and Garnaby (2015), however, could not substantiate an effect of peppermint compared to a placebo on car drivers' alertness based on EEG measures. Citrus has been shown to reduce response times in a study by De Wijk and Zijlstra (2012).

The studies of this thesis are investigating the effect of thermal stimulation. This is because the few existing studies from literature on thermal stimulation that were performed in the context of driving yielded inconclusive results. Chapter 2.2.4 will show that the existing studies used different samples, temperatures and cooled body parts, which is why it is hard to see patterns in the conditions under which thermal stimulation is successfully mitigating fatigue. Because of this gap in research, this thesis is dedicated to further investigate thermal stimulation.

In contrast to the studies on thermal stimulation, the above mentioned studies of mental and haptic stimulation are in agreement. This is why more extensive research on mental and haptic stimulation is queued behind. Olfactory stimulation is fairly unexplored as a countermeasure against passive driver fatigue with only the study of Mahachandra and Garnaby (2015) being conducted in a driving context. The investigation of olfaction comes with the difficulty that there is currently no possibility to verify the emotional effect of odors using objective measures and the risk that odors – out of all stimuli – impact the passenger's discomfort the most, if it is perceived as unpleasant (Bubb et al., 2015).

Both the above detailed research gap in the investigation of thermal stimulation, and the comparison with other directions (mental, haptic and olfactory) substantiate the choice to further analyze thermal stimulation.

2.1.3 Measurement of Fatigue

There exist different methods for measuring fatigue. Subjective measures of fatigue are reviewed in Chapter 2.1.3.1 and objective measures are listed in Chapter 2.1.3.2.

2.1.3.1 Subjective Measures

The most common subjective measures for fatigue in driving experiments are the **Karolinska Sleepiness Scale (KSS)** (A2.1) and the **Stanford Sleepiness Scale (SSS)** (A1). Liu et al. (2009) found that in 11 out of 14 cases either the KSS (8 studies) or the SSS (3 studies) was selected by the authors for subjective assessment in driving studies, which were searched using the terms "driver", "drowsiness", "sleepiness" and "fatigue".

The KSS is a 9-point scale, in which every other point is labeled with a verbal anchor. The steps in between have no verbal label, but can also be selected by the respondents (see A2.1, p. 176). The scale was developed by Åkerstedt and Gillberg (1990) for an experiment that investigated the correlations between subjective and objective sleepiness. One positive aspect of the scale is that it can be easily assessed verbally, without paper and pencil, because the verbal anchors are short and they can be read aloud to the study participant by the investigator. The KSS has also been shown to be closely related to EEG and behavioral variables (Kaida et al., 2006; Putilov and Donskaya, 2013), which validates its use in measuring sleepiness.

The SSS is a 7-point scale, which was originally developed by Hoddes et al. (1973) to measure sleepiness in patients with sleep complaints, but now it is one of the most used scales, including in the field of driving studies. Each point of the scale is labeled with several verbal descriptions (see A1, p. 175).

The Visual Analog Scale is another frequently used tool, and the scale ranges from the extremes "very sleepy" to "very alert" or "not at all" and "very much". The participants have to assess their state by setting a mark on a 100-mm long line (Curcio et al., 2001; Monk et al., 1985).

Other scales for measuring sleepiness that are rarely used in driving studies, are:

- Epworth Sleepiness Scale (Johns, 1991)
- Daytime Sleepiness Scale and Nocturnal Sleep Onset Scale (Rosenthal et al., 1993)
- Sleep Wake Activity Inventory (Rosenthal et al., 1993).
- Accumulated Time with Sleepiness Scale (Gillberg et al., 1994)
- Fatigue, Anergy, Consciousness, Energized and Sleepiness Adjective Checklist (Shapiro et al., 2002)

Table 1 provides an overview of the questionnaires and references to studies in which the scales have been validated to measure fatigue. Curcio et al. (2001) and Shahid et al. (2010) provide a comprehensive review of the above scales. Because of the predominant use of the KSS and SSS in driving studies, only these two scales were used for the experiments of this thesis.

Table 1:   Subjective measures for sleepiness and affective states
Source: Own table

Subjective measure	Introduced by	Validated by	Means of validation
Karolinska Sleepiness Scale	Åkerstedt and Gillberg (1990)	Kaida et al. (2006), Putilov and Donskaya (2013)	correlation with EEG measures, number of lapses and reaction times during vigilance task
Stanford Sleepiness Scale	Hoddes et al. (1973)	Herscovitch and Broughton (1981), Maclean et al. (1992)	correlation with vigilance task performance, principal components analysis of 24 items derived from the scale descriptors
Visual Analog Scale	Monk et al. (1985)	Casagrande et al. (1997)	correlation with partial or total sleep deprivation and task performance

Epworth Sleepiness Scale	Johns (1991)	Johns (1992), Olson et al. (1998)	reliability and internal consistency using factor analysis, correlation with mean sleep latency and total sleep time
Daytime Sleepiness Scale and Nocturnal Sleep Onset Scale	Rosenthal et al. (1993)	Johnson et al. (1999)	replicability validation using factor analyses
Sleep Wake Activity Inventory	Rosenthal et al. (1993)	Rosenthal et al. (1993)	forward stepwise regression analysis with the average sleep latency on the Multiple Sleep Latency Test and factor analysis
Accumulated Time with Sleepiness Scale	Gillberg et al. (1994)	Gillberg et al. (1994)	correlation with the KSS, Visual Analog Scale and vigilance task performance
Fatigue, Anergy, Consciousness, Energized and Sleepiness Adjective Checklist	Shapiro et al. (2002)	Shapiro et al. (2002)	correlations with a number of independent indices
Positive and Negative Affect Schedule	Watson et al. (1988)	DePaoli and Sweeney (2000), Watson et al. (1988)	internal consistency reliability and alpha reliability
Activation Deactivation Adjective Checklist	Thayer (1989)	Ekkekakis et al. (2005), Nemanick and Munz (1994)	principal components factor analysis, varimax rotation, product-moment intercorrelation matrices
Affect Grid	Russell et al. (1989)	Holbrook and Gardner (1993), Russell et al. (1989)	convergent validity discriminant validity and reliability validation using correlations of pleasure and arousal scores at various times
Self-Assessment Mannequin	Bradley and Lang (1994)	Backs et al. (2005), Bradley and Lang (1994)	internal consistency and reliability, correlation with Semantic Differential scale

Passive fatigue, as a state of low arousal can also be viewed as an affective state in which individuals feel little activation (Russell, 1980). Both in Russell's (1980) circumplex model of affect, and in the mood theory of Watson and Tellegen (1985), affective states are viewed as combinations of the two basic orthogonal constructs, valence and arousal. This is also the case for semantic differential methods which are tools used for subjective assessment of valence, arousal and dominance that are associated with affective reactions to diverse stimuli (Bradley and Lang, 1994). (Schmidt et al., 2017a)

Examples of questionnaires to measure the affective states are:

- Positive and Negative Affect Schedule (Watson et al., 1988)
- Activation Deactivation Adjective Checklist, ADACL (Thayer, 1989)
- Affect Grid (Russell et al., 1989)
- Self-Assessment Mannequin (Bradley and Lang, 1994)

Validation of the subjective scales to measure affective states is referenced in Table 1. Only the ADACL was utilized in this thesis (see A2.1). The **ADACL is a list of twenty mood adjectives** in which each adjective is rated on a 4-point scale. The adjectives can be allocated to the subscales **"energy" and "tension"** (for activation) as well as **"tiredness" and "calmness"** (for deactivation). Five adjective scores are averaged to assess each of the four subscales. The ADACL was favorable to the other scales because the verbal descriptors of the ADACL matched the driving context better because it does not include descriptors such as "ashamed" or "afraid" as in the Positive and Negative Affect Schedule (Watson et al., 1988) or "depression" as in the Affect Grid (Russell et al., 1989). Besides, the ADACL as a verbal questionnaire was preferred to the pictographic Self-Assessment Mannequin.

2.1.3.2 Objective Measures

There are also objective indicators of fatigue such as physiological data or performance measures from vigilance tasks (Schmidtke, 1965). Table 2 provides an overview of the physiological indicators of driver fatigue. Most important among these is the **EEG measurement**. The EEG alpha wave power that stems from the alpha brain waves, which are neural oscillations in a certain frequency band, has been shown to correlate with the number of errors in a vigilance test (Kaida et al., 2007). In addition, other frequency bands in the EEG signal (beta, theta and delta) are correlated to fatigue. In the driving simulator study of Jagannath and Balasubramanian (2014) a significant increase of alpha activity, a decrease of beta activity and an increase of theta activity was found. Lal and Craig (2002) found in driving simulator studies that theta and delta activity significantly increased during fatigue. Lal and Craig (2001) provided a detailed overview about EEG studies on driver fatigue.

Indications of fatigue can also be found in **ECG measures**. HR has been reported by Lal and Craig (2000) to decrease significantly during fatigue. The HR reflects the interplay between the sympathetic nervous system (SNS) with the parasympathetic nervous system (PNS), which are both parts of the autonomic nervous system (ANS) (Sammito et al., 2016). The sympathetic branch is responsible for activated states, e.g., for the fight-or-flight response when confronted with danger. The PNS is responsible for relaxed states (Pattyn et al., 2008). The ANS releases a chemical with its parasympathetic branch which leads to an increase in the variability of HR (Sammito et al., 2016). This **increased heart rate variability (HRV)** can be observed in driver fatigue research.

There are several measures that describe the variability in the inter-beat intervals of the HR signal. So-called time domain measures are the standard deviation of normal-to-normal intervals (SDNN) and the root mean square of successive differences (RMSSD) of intervals. The RMSSD is influenced by the parasympathetic nervous system (Sammito et al., 2016). The study of Kaida et al. (2007) revealed that SDNN and RMSSD were correlated with fatigue and predicted fatigue, respectively.

Table 2:   Physiological indicators of driver fatigue
Source: Own table

Signal	During fatigue	References
EEG alpha band power	↗	Curcio et al. 2001, Jagannath and Balasubramanian (2014), Kaida et al. (2007)
EEG beta band power	↘	Jagannath and Balasubramanian (2014)
EEG theta band power	↗	Jagannath and Balasubramanian (2014), Lal and Craig (2002), Putilov and Donskaya (2013)
EEG delta band power	↗	Lal and Craig (2002)
HR	↘	Lal and Craig (2000)
SDNN	↗	Kaida et al. (2007)
RMSSD	↗	Kaida et al. (2007)
LF	↗	Eglund (1982), Michail et al. (2008), Tran et al. (2009), Verwey and Zaidel (1999), Zhao et al. (2012)
HF	↘	Zhao et al. (2012)
LF/HF	↗ ↘	Tran et al. (2009), Wang et al. (2017) Michail et al. (2008), Patel et al. (2011)
PERCLOS	↗	Jarosch et al. (2017)
blink frequency	↗	Körber et al. (2015), Verwey and Zaidel (2000)
blink duration	↗	Körber et al. (2015)
eye movement speed	↘	Lal and Craig (2000)
pupil diameter	↘	Körber et al. (2015)
SCL	↘	Kim et al. (2013)

Other HRV measures can be retrieved from a spectral analysis of the inter-beat intervals. The LF (low frequency component) reflects the oscillations that lie in the frequency band of 0.04-0.15 Hz (Hertz), and the HF (high frequency component) in the band of 0.15-0.4 Hz, respectively. The HF is regulated by parasympathetic activity, whereas the LF is influenced by both the SNS and the PNS (Sammito et al., 2016). Driving studies showed an increase in LF (Eglund, 1982; Michail et al., 2008) and in the ratio LF/HF (Tran et al., 2009; Wang et al., 2017) during fatigue, while others have shown a decrease in LF/HF (Michail et al., 2008; Patel et al., 2011). The inconsistencies in the LF/HF reactions may be due to the calculation of the parameters, which is affected by differently chosen sampling frequencies and the window used for the Fourier transform. Another explanation could be the different study designs.

In addition to EEG and ECG, other physiological indicators of fatigue are eye related parameters. Among those are the **PERCLOS, blink frequency and blink duration** (see Table 2), which have all been shown to increase during fatigue. In a study of Lal and Craig (2000), fast **eye movements** were observed when the driver was alert, compared to limited eye movements in a fatigued state. Körber et al. (2015) found a decrease in **pupil diameter** caused by monotony-induced fatigue. In general, the pupil diameter is regulated by the ANS, and reflects changes in emotional arousal (Partala and Surakka, 2003).

Similarly, the **skin conductance level (SCL)** has been reported as an autonomic measure (Kreibig, 2010) and the study of Kim et al. (2013) provided indication that SCL was correlated with visual fatigue.

The selection of physiological signals that were recorded in the studies of this thesis is described in Chapter 3.6. The studies did not include vigilance tasks such as in previous studies (Casagrande et al., 1997; Herscovitch and Broughton, 1981; Kaida et al., 2006) because this would have altered the task load of the participants and interfered with the otherwise monotonous driving setting.

2.1.4 Reported Models for Fatigue Detection

A frequently used tool in the field of driver and passenger ergonomics is digital human modeling (Bullinger-Hoffmann and Mühlsted, 2016). Human models can be differentiated in anthropometric and cognitive models. Cognitive models are used to predict the behavior of the interaction between driver and vehicle (Bubb et al., 2015).

In the past decade, research efforts have been made to **model fatigue of the driver** in the form of classification algorithms. These algorithms make use of the above physiological indicators of fatigue and the methods of pattern recognition. A list of fatigue models is given in Table 3. Patel et al.

(2011) for example, described a neural network classifying sleepiness with 90 % accuracy based on drivers' ECG data. Friedrichs and Yang (2010) and Hu and Zheng (2009) reported accuracies of 83 % when differentiating three degrees of sleepiness based on camera or EOG (electrooculogram) data extracting eye features. These algorithms have been developed using data from sleep-deprived drivers, hence these are detecting SR fatigue. The model of Igasaki et al. (2015) used data of non-sleep deprived drivers. Their logistic regression based on HRV measures and respiratory features yielded 81 % detection accuracy, however, it was only generated with data from eight male drivers. (Schmidt et al., 2017b)

Karrer et al. (2004) have investigated how individual differences may affect fatigue predictions based on driving behavior and eyelid parameters. They proposed that a classification of people with different fatigue characteristics should be included in fatigue prediction models in order to increase the prediction performance.

Table 3:   Overview of proposed classifiers for driver fatigue
            (SVM, support vector machine; LDA, linear discriminant analysis)

Source: Own table

Reference	ECG	EEG	SCL	Eye data	Breathing	Head or facial data	Driving data	Method	Accuracy	Sample
Bundele and Banerjee (2009)		x						SVM and neural network	96 % and 93 % for two classes	10 professional drivers
Friedrichs and Yang (2010)				x		x		neural network	83 % for three classes	70 % of data were of sleep-deprived drivers
Hu and Zheng (2009)				x				SVM	83 % for three classes	37 sleep-deprived drivers

Reference					Method	Accuracy	Subjects
Igasaki et al. (2015)	x		x		logistic regression	81 % for two classes	8 non-sleep-deprived male drivers
Khushaba et al. (2011)	x	x	x		Kernel-based LDA	97 % for five classes	31 non-sleep-deprived drivers
Krajewski et al. (2009)				x	combination of SVM and k-nearest neighbor	86 % for two classes	12 sleep-deprived drivers
Li and Chung (2013)	x				SVM	95 % for two classes	4 drivers
Murata et al. (2015)				x	multinomial regression	90 % for two classes	13 sleep-deprived male drivers
Patel et al. (2011)	x				neural network	90 % for two classes	12 sleep-deprived professional drivers
Sandberg et al. (2011)				x	neural network	83 % for two classes	12 sleep-deprived drivers
Sayed and Eskandarian (2001)				x	neural network	90 % for two classes	12 partly sleep-deprived drivers
Vicente et al. (2011)	x				LDA	94 % for two classes	11 sleep-deprived drivers

There is a trend to develop in-car fatigue detection systems without the use of driving related parameters. This is because of the movement towards autonomous driving. In SAE (Society of Automotive Engineers) level 3 (SAE Standard J3016), the vehicle is performing the steering, acceleration and deceleration, as well as the monitoring of the environment and therefore, driving data cannot serve as the driver's behavioral measures. At the same time, in SAE level 3 (SAE Standard J3016), the driver is still the fallback in case of a takeover request. Because of this fallback function, it is still essential to monitor the driver's state, including fatigue, with parameters other than driving data.

This literature review on fatigue modeling showed, that many researchers have proposed algorithms to assess fatigue of drivers, but there is currently **no prevalent algorithm** to which researchers refer. From the above overview in Table 3 it can be concluded that most fatigue models were trained using data from sleep-deprived drivers and that the sample sizes are often fewer than 15 people.

2.2 Effect of Thermal Environment on Passive Fatigue

Several theories and studies about the effect of the thermal environment on the driver's state are reviewed in this chapter. First, a little background to the basics of human thermal environments is required to lay the theoretical foundation for the effect of thermal stimulation (Chapter 2.2.1). This includes the underlying neurological processes in thermal perception.

Next, several studies are reviewed that deal with the effect of thermal stimulation. During the review of research investigating thermal stimulation, three different kinds of study settings stand out: interview studies, laboratory studies and driving studies. Therefore, the following chapters are structured accordingly.

Chapter 2.2.2 reviews interview studies of drivers in which common countermeasures against fatigue are retrieved. While these studies delivered valuable input about the subjective perceptions of drivers on cooling, they did not provide objective measures to substantiate the effect. The next Chapter 2.2.3 deals with laboratory studies that investigated thermal stimulation of different body parts in more or less clinical settings. On the one hand, these studies provided physiological effect analyses of the stimulus along with detailed descriptions of the activated neurological pathways during the process. On the other hand, the studies neither included a subjective assessment of fatigue, nor were the participants required to perform a driving task. The last Chapter 2.2.4, gives an overview of the few existing studies about thermal stimulation in a driving context.

2.2.1 Basics of Thermal Perception and Physiological Effects

Thermal perception of humans is enabled through temperature sensors, which are located in the hypothalamus, which is a part of the nervous system, spinal cord and other body parts. Two types of thermal sensors are distributed over the skin, warm and cold receptors. Nervous system pathways link these thermosensors to the hypothalamus. Based on the sensor input, different parts of the hypothalamus are responsible for controlling heat loss, e.g., through sweating, and for heat preservation, e.g., through vasoconstriction (Parsons, 2002). The **density of cold receptors is high in the lips, face, abdomen and chest** (Hensel, 1981). The receptors are sparsely distributed in hands and legs. The distribution of cold receptors in these body parts is given in Table 4.

Table 4:   Distribution of cold receptors in different body parts
Source: *Hensel, 1981*

Body part	Number of cold receptors per cm^2
lips	16-19
face	9
abdomen	8-13
chest	9-10
hands	1-7
legs	3-6

Whereas cold sensation is dictated by mean skin temperature, warm sensation is only initially dependent on skin temperature, then on deep body temperature (McIntyre, 1980). Moreover, Kato et al. (2001) suggested that thermal sensation depends on skin temperature, and that thermal comfort depends on both the skin and core temperatures.

The four basic environmental factors that define the thermal perception of the environment are:

- air temperature
- radiant temperature
- humidity
- air movement

The human factors are:

- metabolic rate
- clothing

In combination, environmental and human factors define the human thermal environment (Parsons, 2002).

There is a theoretical model postulating that ambient temperature influences arousal (Parsons, 2002; Wyon, 1973). The least arousal is perceived at comfortable temperatures (see Figure 4). Increases or decreases in temperature towards less comfortable ranges results in arousal growing steadily.

The Yerkes-Dodson Law describes an inverted u-shaped relation between arousal and performance (Yerkes and Dodson, 1908). According to this law, human performance is weak when the arousal is low, e.g., when being deeply relaxed. On the contrary, very high arousal, e.g., when being anxious, results in deteriorating performance as well. This means "that there is an optimal level of arousal for an optimal performance" (Cohen, 2011, p. 203). Hygge (1992) found also that this optimum level of arousal depends on the difficulty of tasks. Less demanding tasks require higher levels of arousal than difficult tasks (Figure 4).

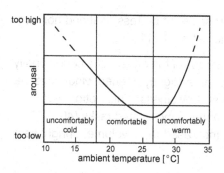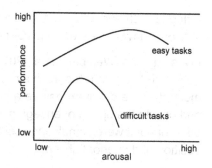

Figure 4:   Relationship between temperature and arousal as well as performance and arousal

Source: Hygge, 1992; Parsons, 2002; Wyon, 1973

Now, these two theories combined suggest that arousal and performance are influenced by the thermal environment and that easy tasks require higher levels of arousal for best performance. Because of this relationship, it may be concluded that **thermal stimulation increases arousal** hence reducing passive fatigue and increasing performance to a certain point. Tejero Gimeno et al. (2006) stated furthermore, that the temperature is an exogenous factor of driver fatigue development with higher temperatures provoking drowsiness and reduced alertness.

2.2.2 Interview Studies of Countermeasures for Fatigue

Interview studies surveying common practice countermeasures against fatigue reported that **"opening a window" or "turning on the AC"** are among the most frequently used countermeasures (Schmidt and Bullinger, 2019).

Anund et al. (2008) found that as a countermeasure against fatigue, 47 % of interviewees open the window and 16 % turn on the AC. In the national survey of Royal (2003), 26 % of the respondents indicated to combat sleepiness by opening a window and it was the second most common named countermeasure after pulling over to rest. Opening the window was also rated as one of the most effective interventions with a mean effectiveness rating of 3.25 and 3.22 on a 1-5 Likert scale for professional and non-professional drivers, respectively (Gershon et al., 2011).

In a study from Weinbeer et al. (2017), 33 % of the drivers of a conditionally automated vehicle rated fresh air through a slightly opened window as the most effective countermeasure against drowsiness. This option was second only to the option of engaging in non-driving related tasks (40 %). The other options were stimulating scent, increased radio volume, upright seat position and interior lighting.

Oron-Gilad and Shinar (2000) recorded similar results on the usage of opening the window and perceived effectiveness. Besides window opening, they found that both civilians and military truck drivers also perceive washing the face as one of the most effective countermeasures.

Pylkkönen et al. (2015) furthermore proposed that there is a dependency on time of day, when alertness-enhancing activities, such as turning on the AC, are used by professional drivers.

These results suggest that **cool and fresh air are subjectively perceived to mitigate fatigue**. The interview studies, however, lack objective evidence for this perception, which is why the review of related work is extended to controlled laboratory and driving studies in the following chapters.

2.2.3 Laboratory Studies of the Effects of Thermal Stimulation

Laboratory studies of thermal stimuli support the suggestion from the surveys of an awakening effect of thermal stimulation (see Table 5). In these studies, a stimulus (e.g. cold water or cold air) has been applied to different body parts and physiological data has been recorded. An initial overview of these studies was provided by Van Veen et al. (2014) in their review about measures to enhance driver vigilance. This thesis adds on to this list in Table 5. Many of the studies in Table 5 suggested that an activation of the SNS is initiated by the stimulus, as indicated by HR reaction patterns (Collins et al., 1996; Hayward et al., 1976; Heath and Downey, 1990; Heindl et al., 2004; LeBlanc et al., 1975, 1976; Sendowski et al., 2000).

The overview in Table 5 shows that thermal stimulation yields physiological responses that can be associated with an activation of the SNS. In addition, it has been shown, that the measured effects occur after as little as one minute for very cold stimuli (Collins et al., 1996; Hayward et al., 1976; Heath and Downey, 1990; Heindl et al., 2004). These **laboratory studies confirm the physiological effects of thermal stimulation** in clinical settings, however, they have been performed outside the context of operator fatigue and therefore subjective ratings of sleepiness have not been recorded. Furthermore, the participants of these studies were for the most part, not performing tasks, as for example driving a car, which is why further literature review was necessary to understand effects of thermal stimuli while driving.

Table 5:   Reported effects of thermal stimulation
Source: Adopted from Van Veen et al. (2014)

Reference	Body part	Stimulus	Time to effect	Effect description
Collins et al. (1996)	face	3 °C cold wind	1 minute	HR decrease, systolic and arterial blood pressure increase
Hayward et al. (1976)	face, abdomen	0 °C to 10 °C cold wind	immediate	HR decrease, sympathetically mediated vasoconstriction (for face cooling only)
Heath and Downey (1990)	face	0 °C to 15 °C cold compresses	immediate	HR decrease, BF (breathing frequency) decrease, blood flow to the finger, toe and calf, respectively
Heindl et al. (2004)	forehead, mouth, nose, back of hand	0 °C ice cubes	1 minute	SNS activation, blood pressure increase
Janský et al. (2003)	lower legs	12 °C water bath	3-10 minutes	SNS activation, HR increase, blood pressure increase
Kawahara et al. (1989)	whole body	12 °C room temperature	not reported	SNS activation, pulse rate decrease, systolic and diastolic blood pressure increase

Koehn et al. (2012)	head and neck	4 °C gel-hat device	immediate	blood pressure increase
Kregel et al. (1992)	one hand	0 °C and 7 °C water bath	1 minute	SNS activation, blood pressure increase
LeBlanc et al. (1975)	one hand, face	5 °C water bath	1 minute	SNS activation and HR increase for hand cooling, SNS activation and HR decrease for face cooling
LeBlanc et al. (1976)	face	-15 °C to 15 °C cold wind	1 minute	HR decrease
LeBlanc et al. (1978)	one hand, face	5 °C water bath, 0 °C cold wind	1 minute	HR increase and systolic blood pressure increase for hand cooling, HR decrease and systolic blood pressure increase for face cooling
Lossius et al. (1994)	whole body	18 °C room temperature	not reported	HRV (LF) decrease, blood pressure variability decrease
Sendowski et al. (2000)	right hand, right index finger, left hand with right index finger	5 °C water bath	<5 minutes	SNS activation, HR increase, blood pressure increase, noradrenaline concentration increase
Tham and Willem (2010)	whole body	20 °C room temperature	continuous effects	SNS activation, increase of α-amylase level

2.2.4 Driving Studies of the Effect of Climate on Fatigue

Schmidt and Bullinger (2019) reviewed the few driving studies, which investigated the effect of a thermal stimulus and the **variety of different results** can be seen in Table 6.

Table 6:   Studies investigating thermal stimulation to mitigate driver fatigue
Source: Adopted from Schmidt and Bullinger (2019)

Reference	Stimulus temperature	Stimulus duration	Body part	Sample	Effect of stimulus
Landström et al. (1999)	18 °C	2 minutes and 4 minutes	head	sleep-deprived	significant increase in subjective wakefulness, significant decrease in alpha activity (EEG)
Landström et al. (2002)	temperature drops of 8 °C to 10 °C from 25 °C to 30 °C	2, 4, 6 and 8 minutes	not specified	sleep-deprived	significant increase in subjective wakefulness
Reyner and Horne (1998)	10 °C	2 hours	face	sleep-deprived	non-significant decrease in subjective sleepiness, non-significant difference in EEG power (4-11 Hz)
Schwarz et al. (2012)	not specified	10 minutes	not specified	non-sleep-deprived	non-significant decrease subjective sleepiness and blink duration
Van Veen (2016)	temperature drops of 5 °C	4 minutes	hands	non-sleep-deprived	non-significant decrease in subjective drowsiness, significant increase in HR
Wyon et al. (1996)	21 °C	1 hour	compartment temperature	non-sleep-deprived	better task performance

Landström et al. (1999) investigated the effect of short-term room temperature drops from 28 °C to 18 °C on subjective drowsiness and EEG data. The results showed a significant decrease in subjective sleepiness in the cooling condition and EEG data supported an enhanced wakefulness. Sleep-deprived subjects were recruited for this study. In a later field study, Landström et al. (2002) equipped trucks with a temperature control system which cooled down the cabin from starting temperatures between 25 °C and 30 °C repeatedly by 8 °C to 10 °C. They reported an increased alertness of the professional drivers when using the system. One of the causal factors for the drivers' fatigue in this study was sleep deprivation because the truck drivers used the temperature control system mainly during the night and after long hours of driving.

Reyner and Horne (1998) showed through a simulator study, that a treatment of 10 °C cold air for the period of two hours neither had a significant effect on subjective sleepiness ratings nor on EEG data. Their study also focused on drowsiness caused by sleep deprivation.

Schwarz et al. (2012) recruited subjects which were not sleep-deprived and researched the effect of intermittent 10-minute periods of opening the window by 2 cm at speeds of 120 km/h in a real-driving study. No temperature measurements were reported and subjective sleepiness and blink duration were not affected by this countermeasure.

A simulator study, which focused on passive fatigue, was conducted by Van Veen (2016) who investigated the effect of intermittent local hand cooling by a temperature difference of 5 °C to the ambient temperature on non-sleep-deprived subjects. A significant increase of HR after 3 cooling minutes was measured which indicated an activation of the SNS, however, the effect of local cooling of the hands for repeated 4-minute periods on perceived sleepiness was not significant.

In another example, Wyon et al. (1996) found a better task performance in detecting changes in the instrument cluster of a car during a driving study at a temperature of 21 °C compared to 27 °C.

In conclusion, the overview in Table 6 shows that the studies yielded different outcomes in terms of effectiveness of thermal stimulation against passive fatigue. This is due to different stimulus temperatures, durations, cooled body parts and the degree of sleep deprivation in the sample. It remains unclear under which stimulus conditions passive driver fatigue can be mitigated, especially while maintaining the driver's comfort.

2.3 Conclusion

From the review of related work, several conclusions can be drawn. There exist different types of fatigue and their differentiation is crucial as the different types can be attributed to **different causal factors** and importantly, **different countermeasures are needed** to mitigate these types of fatigue.

Passive fatigue gains more and more relevance in the context of increasing automation which led to the decision to focus on passive fatigue for the empirical studies of this thesis. A detailed review of studies investigating causes and countermeasures of passive fatigue revealed several insights. First, diverse causes for passive fatigue were identified that can be clustered into **environmental factors, vehicular factors and personal factors** (Figure 3). Second, the countermeasures can be clustered into **mental, haptic, thermal and olfactory stimulation**. Previous studies on mental and haptic stimulation are in agreement in terms of the yielded effects, which is why more extensive research on these matters is queued behind. So is olfactory stimulation because there is currently no possibility to verify the emotional effect of odors using objective measures and the risk that odors affect the passenger's discomfort the most, if it is perceived as unpleasant (Bubb et al., 2015). This leaves the **fairly unexplored field of thermal stimulation** as a countermeasure against passive fatigue in the driving context, for which it remains unclear under which stimulus conditions passive fatigue can be reduced. This research gap will be addressed by the studies in Chapter 5.

The literature review showed that the **KSS and SSS are the most frequently used** subjective measures for fatigue in driving studies (according to Liu et al. (2009)), which is why these measures are retrieved in the studies of this thesis. There is also a variety of physiological signals that have been shown to be correlated with fatigue (Table 2).

The review of fatigue models using objective input parameters to model fatigue revealed that there is no prevalent model that is commonly used among researchers. Instead, several researchers have proposed their own model that has been trained using data from fairly small samples, with the majority being sleep-deprived (Table 3). The **lack of a prevalent fatigue**

model led to the decision to generate a new model in Chapter 6 using data from a large sample (n=88) of non-sleep-deprived drivers.

Interview studies confirmed that turning on the AC or opening a window is a frequently used countermeasure against fatigue and perceived as effective by professional and non-professional drivers. The interview studies, however, lack objective evidence for this perception. Laboratory studies outside the driving context (Table 5) showed that thermal stimulation yields physiological responses that can be associated with an activation of the SNS. Lastly, the few driving studies applying thermal stimulation as a countermeasure against fatigue yielded **different outcomes in terms of effectiveness of thermal stimulation** against passive fatigue, due to different stimulus and sample characteristics. To explore under which stimulus conditions passive driver fatigue can be mitigated, several studies are performed (see Chapter 5).

3 Tools and Methods[1]

To research thermal stimulation as a countermeasure against passive fatigue, several studies have been conducted. This chapter provides details about the applied tools and methods in these studies. Starting in Chapter 3.1, the overall design of the different studies and their relation are explained. Details about the driving simulators and vehicular apparatuses are provided in Chapters 3.2 and 3.3. The different thermal stimuli that were tested in the studies are described in Chapter 3.4. Chapter 3.5 deals with the recruitment method and characteristics of the study participants. The rationale for recording specific physiological measurements and information on the sensors used are provided in Chapter 3.6. The chapter closes with an overview of the data analysis (Chapter 3.7).

3.1 Study Design

As explained in Chapter 1.2, the first RQ addressed the feasibility of inducing passive fatigue for experimental purposes. To answer this question, a pre-study was necessary in which a driving scenario was tested with respect to its **capability to induce fatigue in drivers without sleep deprivation**, solely by the monotonous nature of the drive. To measure the fatigue of drivers and hence, the suitability of the drive for the following studies, subjective and objective indicators of fatigue were measured.

---

[1] The description, layout and photos of the simulator, apparatus, physiological measurements and data analysis are based on a previous publication: Schmidt, E., Decke, R., Rasshofer, R., & Bullinger, A. C. (2017a). Psychophysiological responses to short-term cooling during a simulated monotonous driving task. *Applied Ergonomics, 62*, 9-18. https://doi.org/10.1016/j.apergo.2017.01.017.

The second RQ addressed the effect of different thermal stimuli on fatigued drivers. Studies to investigate this question required the induction of fatigue in non-sleep-deprived participants by means of monotony. Moreover, **subjective and objective indicators of fatigue** needed to be measured in order to compare different thermal settings. These studies obviously necessitated a way to **control the car climate** in order to provide thermal stimulation.

Consequently, a secondary data analysis of several driving studies was performed to recognize patterns in the physiological data of fatigued drivers. Through a statistical approach, a regression model was developed that detected passive fatigue. The third RQ, namely quantifying the level of accuracy of the model, could be answered by comparing the regression model outcome with the subjective assessments of the drivers. The regression could be used in the following studies for both the triggering of the thermal stimulus and the effect measurements of different stimuli. To generate the fatigue model, a sufficiently large dataset was required that consisted of predictor variables and classes. The predictor variables were various signals such as ECG, skin conductance and pupil diameter. The classes were formed by the subjective fatigue ratings of the driver.

The combination of fatigue detection and intervention was then tested in a final study. As above, the closed-loop study required the induction of fatigue as well as its subjective and objective measurement.

From the above descriptions, three basic requirements could be defined:

- the induction passive fatigue
- the possibility to cool different parts of a car cabin
- the subjective and objective measurement of fatigue

The first requirement on the experiments, fatigue induction, led to the decision to **conduct the studies in driving simulators**. As the studies of Philip et al. (2005), Hallvig et al. (2013) and Fors et al. (2016) have shown, perceived sleepiness and physiological sleepiness are higher in simulated driving than in real driving for both sleep-deprived and non-sleep-deprived subjects due to lower levels of visuomotor stimulation.

Since this study required monotony induced fatigue in non-sleep-deprived subjects, simulator studies were well suited and allowed also for reproducible traffic scenarios.

The second requirement, cooling of the car, was addressed by using either an **external AC attached to the car** or the **car's internal AC**. Which of the two options were used depended on the driving simulator in which the study took place and the required temperature. The minimum temperature of the external AC and the car's AC was 17 °C and 9 °C, respectively. Two different simulators were used for the various driving simulation studies. The reason for not using the same simulator in all studies was due to the fact that strict security regulation prevented the operation of a high voltage battery in the simulator area used for the first study. Therefore, all studies that required colder air from the car's internal AC had to be moved to a simulator with more relaxed facility regulations. Details follow in Chapters 3.2 and 3.3.

To fulfill the last requirement, subjective and objective measurement of fatigue, fatigue was **evaluated with questionnaires and with physiological measurements**. Fatigue was assessed several times during the simulated drives with the SSS (Chapter 4) and KSS (Chapters 5.1, 5.2, 5.3 and 7). Several physiological signals were recorded continuously throughout the drive. Details follow in Chapter 3.6.

3.2 Driving Simulators

The first four studies (Chapter 4, 5.1 and Chapter 6) were conducted in a static driving simulator with a 3.45 m high, curved (Ø 6.30 m) screen providing a 220° view (Figure 5). In this simulator, two monitors with a screen diagonal of 1.27 m facing the side mirrors were placed behind the car to enable a rear view.

Figure 5: BMW i3 in the curved-screen driving simulator
Source: Schmidt et al., 2017a (in the supplemental material)

In the curved-screen simulator, the investigator controlled the experiment in a separate room, which is shown in Figure 6.

Figure 6:   Room for investigator in the curved-screen simulator
Source: Schmidt et al., 2017a (in the supplemental material)

The second simulator was used in the three experiments described in Chapters 5.2, 5.3 and 7. In this simulator, the driving scene was projected on a 3.2 by 2.5 m flat screen. A curtain separated this flat-screen simulator and the investigator's desk as shown in Figure 7.

Figure 7:   Flat-screen simulator with desk for investigator
Source: *Own photograph*

Both simulators recorded the driving parameters at 100 Hz: steering angle, speed, acceleration, gas and brake pedal angle and offset from the middle of the lane. A difference between the simulators was that the flat-screen simulator did not provide a rear view via the side mirrors, but only via a rear mirror display. This difference should not influence the results as the simulated highway traffic scenario chosen in the main studies of this thesis was very light on traffic and therefore the participants were not required to use the side mirrors. Previous dissertations in the field of human factors comprising various studies have also used different driving simulators (Hergeth, 2016; Platten, 2012; Schmidt, 2018).

In general, the driving simulation in this research was used as a means to induce passive fatigue in the context of a driving task rather than to investigate the effect of thermal stimulation on the driving parameters. Instead, the analysis of the effect of thermal stimulation on passive fatigue was performed using subjective and physiological measurements as suggested by previous research (Chapter 2.1.3).

3.3 Apparatus

During the course of this research project different apparatuses were used for the studies.

For the first study described in Chapter 4, a mock-up of a BMW 530 series was used which is shown in Figure 8. This mock-up was only utilized in the curved-screen simulator. When the BMW 530 mock-up was used, a third monitor with a screen diagonal of 1.27 m facing the rear mirror was placed in the middle behind the mock-up to extend the rear view from the side mirrors to the rear mirror.

Figure 8:   Mock-up of a BMW 530 series placed in the curved-screen simulator
Source: *Own photograph*

In all other studies, a street legal BMW i3 was used (Figure 5 and Figure 7). The car was changed only for experimental purposes. The car was equipped with a rear mirror display. The original steering wheel, gas pedal and brake pedal were replaced by corresponding parts from Logitech (Switzerland) to enable the connection to both simulator environments.

Because of strict security regulations in the curved-screen simulator, it was not possible to run the first cooling study (Chapter 5.1) with the high voltage battery of the modified electric vehicle. Therefore, the high voltage battery was removed and consequently the original AC system of the car could not be operated. To provide cold air, an external AC unit was placed behind the screen and the cold air duct was attached to the fresh air intake of the vehicle (Figure 9). Via remote control of the circulation flap and the interior fans of the vehicle, the cold air could be let in the cabin as required without any actions required from the driver. The remote control was operated by the investigator in the control room (Figure 6).

During all experiments with the BMW i3 the trunk and the passenger window stayed open to avoid an increase of CO_2 in the vehicle cabin. In the regular operation mode of the car's AC, the system balances automatically the fresh airflow and the recirculated airflow, which is why in real driving environments, the CO_2 level in the cabin will not increase. In the cooling studies, the AC system of the car was modified such that the regular operation mode was not usable. Therefore, the CO_2 level in the cabin would have been increasing from a basic value of 500 ppm to 3000 ppm over a half hour period of driving with shut windows and doors, as pre-studies have shown. To avoid the influence of a rising CO_2 level on the participants' fatigue, the trunk and the passenger window stayed open and thus the CO_2 level stayed constant.

In order to monitor the vehicle climate and CO_2 levels, the car was equipped with a temperature, relative humidity and CO_2 sensor, mounted between the driver's and passenger's seat at a distance of 30 cm from the ceiling. Another sensor measuring temperature was mounted to the front left air inlet in order to verify similar temperature conditions between experiments. The car's temperature display was disabled so that the drivers would not be visually tipped off to a change in the temperature setting.

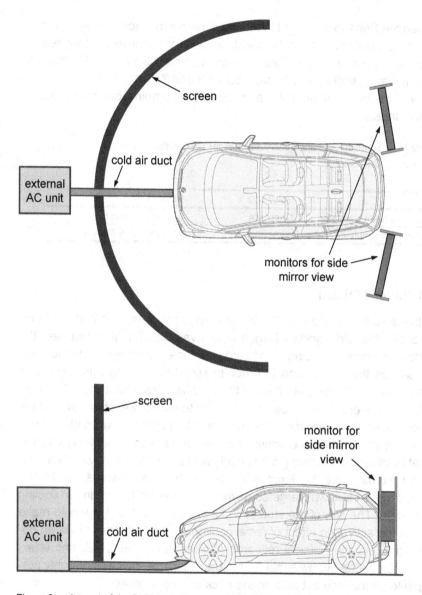

Figure 9:   Layout of the BMW i3 in the curved-screen simulator
Source: *Schmidt et al., 2017a (in the supplemental material)*

When the BMW i3 was used in the flat-screen simulator (Figure 7), the car could be operated with its high voltage battery because of more relaxed security regulations. Therefore, the car's original AC was used in the cooling studies described in Chapters 5.2, 5.3 and 7. Table 7 shows an overview of the different simulator and apparatus combinations of the studies of this thesis.

Table 7:   Simulator and apparatus combinations used in the thesis
Source: Own table

	BMW 530	BMW i3
curved-screen simulator	Chapter 4	Chapter 5.1 and 6
flat-screen simulator		Chapter 5.2, 5.3 and 7

3.4 Thermal Stimuli

In the studies of Chapter 5.1, 5.2 and 7, upper body cooling and its effects on subjective and objective fatigue were investigated. In these studies, the middle air vents were directed towards the face of the driver. The face was chosen as the target area in order to stimulate the trigeminal nerve, a mechanism that was described in the cold face test (Heath and Downey, 1990). In general, facial receptors respond to cold or wet stimulation of the area around the nose and eyes by stimulating both the sympathetic-vascular smooth muscle pathways and the cardiac-vagal pathway, causing what is known as the diving reflex (Collins et al., 1996; Heath and Downey, 1990). Furthermore, by directing the middle air vents towards the driver's face, exposed skin was targeted as much as possible for maximum cooling sensation. The aim of the studies in Chapter 5.1, 5.2 and 7 was to make use of the activating effect of facial cooling on the SNS that has been observed in previous clinical research (Collins et al., 1996; LeBlanc et al., 1976). Since the airflow expanded in the cabin, the chill was not only perceptible on the face but also on the neck and upper chest.

In order to explore other methods of thermal stimulation inside a car cabin a study on leg cooling was conducted (Chapter 5.3). This was inspired by water treading and its physiological effects. The term water treading is used in this thesis to refer to a type of hydrotherapy developed by Sebastian Kneipp in the 19th century (Brüggemann, 1980), during which people step into a knee-deep, cold water pool and walk in it or step in place for about up to one minute (Ganz, 2016). The cold stimulus evokes vasoconstriction, followed by vasodilation after the activity, which encourages the blood flow in the skin. This in turn increases the metabolic functions and relieves the cardiovascular system (Ganz, 2016). The literature support on the effects of leg cooling is much smaller compared to the above facial cooling (see Table 5 in Chapter 2.2.3), however, Janský et al. (2003) could find patterns of SNS activation and an increase in HR and blood pressure while applying a cold stimulus to the lower legs.

3.5 Participants

The study participants in all studies were BMW Group employees who were recruited via email lists and voluntarily took part in the studies. Recruiting the company's employees means that convenience sampling was applied, i.e., including participants that meet certain practical criteria, such as easy accessibility (Etikan et al., 2016). Convenience sampling is the most widely used sampling method in usability testing (Locascio et al., 2016; Nielsen, 1994), but is vulnerable to hidden biases in the sample (Etikan et al., 2016). Therefore, it is important at this point to address how the sample differs from one that would have been randomly selected (Etikan et al., 2016).

The most obvious differences of the sample compared to a random population sample are: nationality, gender and age. The vast majority of participants were German nationals, approximately two thirds of them were male and, on average, younger than 32 years (see Chapters 4.1.3, 5.1.1.3, 5.2.1.3, 5.3.1.3, 6.1.3 and 7.1.3). Other differences to a random sample could be brand loyalty and health. How these differences of the convenience samples may limit the studies' validity is discussed in detail in Chapter 8.2. Recruiting the company's employees has been applied previously within doctoral research projects in the field of human factors (Hergeth, 2016; Platten, 2012; Schmidt, 2018).

The participants kept their regular sleeping schedule, but were instructed to avoid tobacco, caffeinated beverages and alcohol on the day of the study to eliminate the influence of other stimulants on the physiological behavior. They were asked before the study to wear underwear, socks, shoes, pants and T-shirts. Participants knew ahead of time that physiological data would be recorded. They did not know that cooling would be applied during the experiments and were recruited under the study title "Fatigue detection based on physiological measures".

At the beginning of the experiments, all participants signed a consent form and completed a questionnaire on demographic data, which included questions about the intake of caffeine, alcohol or cigarettes as well as the amount of sleep. They were also asked, whether they feel healthy and willingness to participate in the study.

During the time of the initial questionnaires, attachment and calibration of all sensors, the baseline recording of the ECG and the familiarization drive, participants could acclimate for 30 minutes to the thermal conditions of the simulator before the start of the first test drive.

The participants were examined at different times of the day. A typical test day lasted from 8 am to 6 pm. They main motivation to run the tests throughout an entire workday is related to an efficient use of the booked simulator capacity. Potential drawbacks of all-day testing is that some participants drove during their circadian afternoon dip which could influence the fatigue measurements. All-day testing has been applied in various simulator studies investigating fatigue (Jarosch et al., 2017; Schmidt et al., 2016b). A subsequent group comparison of data collected in the study of Chapter 5.1 has shown, that time of day did not affect the participants' fatigue (see 5.1.1.3).

3.6 Physiological Measurements

In all experiments, physiological signals were recorded. Medical sensors of g.tec (Austria) were used for measuring a 3-channel-ECG, BF and SCL.

ECG was recorded because HR, SDNN, RMSSD, LH and HF can be extracted from the ECG signal. As explained in Chapter 2.1.3.2, these signals have been shown to correlate with either subjective fatigue measures or performance in vigilance tasks. HR has been reported by Lal and Craig (2000) to decrease significantly during fatigue. The study of Kaida et al. (2007) revealed that SDNN and RMSSD were correlated with fatigue or predicted fatigue, respectively. Similarly, SCL has been shown by Kim et al. (2013) to be correlated with settings inducing different degrees of fatigue. Last, changes in breathing indicated fatigue with a similarly high accuracy as ECG data (Igasaki et al., 2015). Because of these insights, these signals were measured in all studies and used as objective indicators of fatigue.

These signals were sampled with 512 Hz. The ECG electrodes were placed under the right and left clavicle and on the lower left abdomen within the rib cage frame. The BF was measured by means of an elastic belt, worn around the chest. The electrodes for the SCL were attached to the middle and ring finger of the left hand (Figure 10). In all experiments, a baseline signal of all physiological measures was recorded for 10 minutes with the participants sitting in a resting position in the driver's seat.

Figure 10:   Participant of a driving simulator study equipped with physiological sensors
Source: Own photograph

As described in Chapter 2.1.3.2 (Table 2), eye data, such as pupil diameter, PERCLOS, blink frequency, blink duration and eye movement speed have been reported as useful indicators of fatigue (Jarosch et al., 2017; Körber et al., 2015; Verwey and Zaidel, 2000). Therefore eye tracking parameters were also used as objective measures of fatigue.

In all studies, remote eye trackers were mounted on the car's dashboard. The reason for using remote eye trackers instead of head-mounted eye trackers was due to the focus on drivers' comfort. State-of-the-art head-mounted eye trackers in the years 2015 to 2017 had a noticeable weight (e.g. 69 grams as reported by Stuart et al. (2016)) and needed to be strapped onto the head tightly to fix the cameras' positions. As the studies lasted up to two hours, head-mounted eye trackers were avoided as they might have led to higher discomfort while driving.

In the first experiment (Chapter 4), gaze coordinates were collected using a Tobii REX eye tracker (Tobii, Sweden). In all other experiments, a newer Tobii Pro X2-60 eye tracker (Tobii, Sweden) was used, which also recorded the pupil diameter at 60 Hz. The pupil diameter has been shown to be an indicator of fatigue by Körber et al. (2015). To avoid external influence on the pupil diameter measurement, the light conditions of the driving simulation were kept constant.

The newer Tobii Pro X2-60 eye tracker provides more accurate gaze points and additional outputs compared to the older Tobii REX eye tracker (Tobii, 2018). The older model was only used in the pre-study (Chapter 4) that dealt with the induction of fatigue. No cross-study comparisons were performed between the pre-study and any of the following studies in which the newer Tobii Pro X2-60 eye tracker was used. Therefore, using two different eye trackers does not distort any conclusions drawn.

With the recording of the gaze coordinates, it is possible to assess the eye movement speed which was found to be high in alert drivers, compared to limited eye movement in fatigued drivers (Lal and Craig, 2000). A drawback of the remote eye trackers is that PERCLOS, blink frequency and blink duration cannot be reliably calculated, which would have been good indicators of fatigue (Chapter 2.1.3.2). During cooling with air movement, the draught could affect PERCLOS and blink frequency, as the study of Acosta et al. (1999) on ocular comfort in various environmental conditions has shown.

Nevertheless, facial action units were extracted as an alternative to PER-CLOS at 100 Hz from a frontal camera (IDS Imaging Development Systems, Germany) facing the subjects from the dashboard with a facial expression recognition software (iMotions, Denmark) which is based on the FACS (facial action coding system) of Ekman and Rosenberg (1997). In the FACS, the action unit number 43 indicates closed eyes, which was of interest for objective fatigue evaluation. Evidence values provided by the software were on a logarithmic scale.

As reported in Table 2, EEG would have been another meaningful source of parameters that have been shown to indicate fatigue. In contrast to the measurement of ECG, skin conductance or eye data, no unobtrusive measurement techniques have been developed for EEG recordings. Usually, electrodes need to be placed along the scalp and solely in-ear sensors currently pose an alternative (Goverdovsky et al., 2016). For more direct practical implications, only ECG, SCL and eye data were recorded where unobtrusive technologies currently exist to measure ECG via a car seat (Eilebrecht et al., 2001), skin conductance via the steering wheel (D'Angelo and Lüth, 2011) and eye related parameters via camera systems (Toyota, 2018).

3.7 Data Analysis

The data was processed and analyzed using Matlab 2013b. Upon recording of the physiological measurements, the signals were filtered to remove noise. The ECG signal was filtered with a 0.01 Hz high pass, a 60 Hz low pass and a 50 Hz notch filter. The BF signal was filtered with a 0.01 Hz high pass, a 30 Hz low pass and a 50 Hz notch filter. The SCL was filtered with a 30 Hz low pass and a 50 Hz notch filter. All filters were selected according to the sensor manufacturer's recommendations (G.tec, 2018).

From the ECG recording, the HR and HRV measures were obtained. Both time and frequency domain HRV were extracted because studies of Eglund (1982), Kaida et al. (2007) and Patel et al. (2011) have proven those as indicators of fatigue. The time domain HRV measures SDNN and

RMSSD were calculated for signal windows of 3 minutes. Furthermore, spectral analyses of 3-minute sequences of interbeat intervals were performed and the frequency domain HRV measures LF and HF were obtained.

The raw SCL was processed using the convex optimization approach of Greco et al. (2014). Eye tracking data were processed to provide gaze variability. When using the Tobii Pro X2-60, the pupil diameters of the left and right eye were averaged for each subject.

For the statistical analyses, all continuously recorded data and evaluated signals were averaged for each minute and for each participant (except for Chapter 7 in which 30-second intervals were used). A significance level of p=.05 was used for all statistical tests, unless stated otherwise.

4 Fatigue Induction in Simulated Driving[2]

The induction of passive fatigue in simulated driving environments with non-sleep-deprived drivers is an important requirement for all studies of this thesis dealing either with the detection or with the mitigation of passive fatigue.

This necessity of the induction of passive fatigue in the driver precipitated RQ 1:

> RQ 1: Which degree of passive fatigue can be induced by means of task underload in drivers according to subjective and objective measures?

To answer RQ 1 a study was conducted that aimed to explore a monotonous driving scenario with regard to its suitability to induce passive fatigue. As explained in Chapter 2.1.1 passive fatigue is caused by task underload (Neubauer et al., 2012). Therefore, passive fatigue is more likely to arise on rural highways than on crowded urban streets (Fell and Black, 1997; Horne and Reyner, 1995; Thiffault and Bergeron, 2003).

To get an idea how fatigued simulator drivers can get while driving on a highway with low task demand, a simulator study was conducted. Another aim of this study was to analyze the correlations between subjective and objective driver fatigue measures. The evaluation of the correlations between these measures in this chapter serves as a screening for suitable input parameters for a fatigue model presented in Chapter 6. In order to

---

[2] This chapter is based on a previous publication: Schmidt, E., Decke, R., & Rasshofer, R. (2016a). Correlation between subjective driver state measures and psychophysiological and vehicular data in simulated driving. *Proceedings of the IEEE Intelligent Vehicles Symposium* (pp. 1380-1385). Gothenburg, Sweden: IEEE. https://doi.org/10.1109/IVS.2016.7535570. © 2016 IEEE

© Springer Fachmedien Wiesbaden GmbH, part of Springer Nature 2020
E. Schmidt, *Effects of Thermal Stimulation during Passive Driver Fatigue*,
Gestaltung hybrider Mensch-Maschine-Systeme/Designing Hybrid Societies,
https://doi.org/10.1007/978-3-658-28158-8_4

generate variance in both subjective and objective measures for a meaningful correlation analyses, the study also included highway drives with more challenging traffic scenarios.

The following chapter is continued as follows. First, the method is explained in Chapter 4.1. Second, the results of subjective measures, objective measures and the relation between those two are reported in Chapter 4.2. Subsequently, the results are discussed in Chapter 4.3 and last, the conclusions about fatigue induction via monotonous traffic scenarios and correlations between subjective and objective fatigue measures are drawn in Chapter 4.4.

4.1 Method

The subsequent chapters are structured as follows. Chapter 4.1.1 gives an overview of the technical setup in which the study took place. The study design is explained in Chapter 4.1.2 and the sample is described in Chapter 4.1.3. The dependent variables collected in the study are listed in Chapter 4.1.4. Last, Chapter 4.1.5 outlines how the collected data were analyzed.

4.1.1 Setup

This pre-study was conducted in the curved-screen simulator (see Chapter 3.2) with the BMW 530 mock-up (see Chapter 3.3).

4.1.2 Study Design

The study employed a repeated-measures design. This means that all participants got to experience all test conditions. This method automatically controls for individual variability and requires less participants than between-subject testing (Nielsen, 1994). Each driver was asked to drive in five different conditions, all on a highway, which aimed to evoke various physiological and subjective responses. The spread in the responses is beneficial for the subsequent correlation analyses. The conditions were:

Baseline – In the first 3 minutes of this drive, participants could try out different driving maneuvers in order to get familiar with the driving simulator. After this, participants were tasked for another 3 minutes to follow a car within a given distance (50 m) which was displayed on the screen. In addition, the actual distance between the participant's car and the "follow-me" vehicle was displayed. The "follow-me" vehicle changed speeds slightly and there were no other vehicles.

Easy – During this 8-minute highway drive light traffic was simulated. The "follow-me" vehicle changed speeds four times by 10 km/h at accelerations of 2 m/s². The target distance to the "follow-me" car changed two times from 50 to 30 m.

Provoking – This drive was the same as the "easy" drive with regard to the traffic and the "follow-me" car behavior. The difference was that participants were told at the beginning that their driving performance would be evaluated. Driving performance in this case consisted of the accuracy in following the car in the given target distance and how well drivers were driving in the middle of the lane. Furthermore, they were told that the driving performance was compared to previous drivers and a ranking would show them their current rank each minute. Fake rankings were presented, however, in order to indicate a poor ranking compared to the other drivers.

Stressful – In this 8-minute drive the traffic was much heavier and included uncooperative road users that swerved in between the participant's car and the "follow-me" vehicle, making it impossible to keep the target distance. Other times, it was not possible for the participants to perform necessary overtaking maneuvers in order to match the distance because all lanes were blocked by other vehicles or because of construction zones on the highway with narrow lanes. Furthermore, the subject was honked at when merging into another lane before entering the construction zone to provoke additional stress. The "follow-me" vehicle performed more extreme speed changes by 20 km/h at accelerations of 7 m/s².

Tiring – This drive did not require the "follow-me" task in order to reduce the workload on the drivers. Additionally, there were no other road users and the road side was designed to look monotonous as in the study by Thiffault and Bergeron (2003), with only trees, and did not include any noticeable eye catchers (like factories or cities for example). The participant was tasked not to drive faster than 120 km/h. The drive lasted for 17 minutes.

The conditions "easy", "provoking" and "stressful" were randomized. The tiring drive was excluded from the randomization in order to avoid the induction of fatigue before the drivers completed all more challenging drives and hence tested last.

4.1.3 Participants

Fifty BMW Group employees were recruited in August 2015 via an email list on a voluntary basis. Because of simulator sickness, four participants could not continue with the study after the baseline drive. Therefore, the sample consisted of 46 participants (30 males, 16 females) with a mean age of 30.5 ± 10.7 y (years).

Details about the requirements for participation are described in Chapter 3.5. Participants did not know that the different drives were intended to induce diverse affective states or fatigue.

4.1.4 Dependent Variables

At the end of each driving condition, participants were asked – amongst other things (for details see Schmidt et al., 2016a) – to assess fatigue using the SSS (Hoddes et al., 1973, German version by Mieg, 2006). Details about the recording of the physiological data are described in Chapter 3.6.

4.1.5 Data Analysis

Analyses of the psychophysiological data were performed on the last minute of recordings of each drive. The last minute was used for evaluation because it took time to build up affective states, such as passive fatigue, during the scenarios. Using the last minute of physiological recordings of each condition is a method for ensuring steady state that has been applied in previous studies investigating the induction of psychological states (König et al., 2011). Further information on signal processing can be found in Chapter 3.7.

The dependent variables were tested for differences between the different drives and also for relations between subjective and objective data.

4.2 Results

In the following, the results of the subjective assessments of fatigue at the end of the different drives are presented in Chapter 4.2.1. Chapter 4.2.2 analyzes the results of the physiological measures. Last, Chapter 4.2.3 lists the results of correlation analyses between subjective and objective data.

4.2.1 Subjective Data

The mean SSS rating after the tiring drive was 4.4 (SD=1.1) which lies between "somewhat foggy, let down" and "foggy, losing interest in remaining awake, slowed down" (Hoddes et al., 1973).

A repeated-measures ANOVA confirmed that the drivers' fatigue changed significantly between the conditions (F(3.37, 151.80)=61.16, p<.001 using

Greenhouse-Geisser correction) (see Figure 11). Post-hoc tests with Bonferroni correction showed that there was a significant difference in SSS ratings between the tiring drive and all other drives.

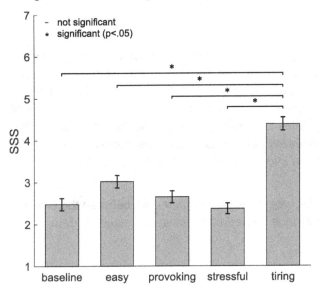

Figure 11:   SSS ratings (M and SE) after the experimental drives
Source: *Adopted from Schmidt et al., 2016a*

4.2.2 Physiological Parameters

Friedman tests of the non-normal data confirmed significant changes of the physiological data (see Table 8). The table also includes Wilcoxon post-hoc test results of Bonferroni-corrected comparisons between the tiring drive and the other conditions.

HR varied significantly between the conditions. As expected, the HR's lowest value was at the end of the tiring drive. The HRV measures SDNN, RMSSD and LF increased significantly in the tiring drive, which is an objective indicator for fatigue. Furthermore, BF was significantly lower in the tiring drive compared to the other drives.

Table 8:   Physiological measures at the end of each drive
            (* significantly different from the tiring drive (p<.05))
Source: Schmidt et al., 2016a, © 2016 IEEE

Signal	Baseline	Easy	Provoking	Stressful	Tiring
HR [bpm]	72.7*	70.4	71.9*	70.6	69.2
SDNN [ms]	44.6*	46.2*	41.5*	49.6*	58.1
RMSSD [ms]	40.0	40.3	37.7*	42.4	44.8
LF [s²]	0.54*	0.55*	0.54*	0.55*	0.59
BF [1/min]	18.9*	18.2*	19.1*	18.6*	16.2

4.2.3 Correlations Between Subjective and Objective Data

The correlation analyses show that few of the physiological signals were moderately correlated with the SSS ratings. The correlation results are shown in Table 9.

Table 9:   Correlations of SSS ratings with physiological data
Source: Schmidt et al., 2016a, © 2016 IEEE

Signal	Correlation with SSS
SDNN	r=0.26, p<.001
BF	r=-0.25, p<.001
standard deviation of the vertical gaze coordinates	r=0.25, p<.001
blink frequency	r=0.26, p<.001

All other parameters that were extracted from the physiological recordings in these studies (see Chapter 3.7) did not yield significant correlations or had a correlation coefficient under |r|=0.2.

4.3 Discussion

The subjective SSS ratings showed that participants on average perceived the onset of fatigue and felt "somewhat foggy, slowed down and lost interest in staying awake" after 17 minutes of monotonous driving. It can be assumed that these perceptions are caused by the task underload experienced in this drive rather than circadian effects because the participants were not sleep-deprived. The degree of perceived fatigue, induced by means of task underload in simulated driving is comparable to the findings of Jarosch et al. (2017) who measured fatigue in a simulator during an automated drive in which participants were required to perform a monotonous monitoring task. After 19 minutes, the average KSS rating was between 6 and 7, whereas a value of 7 is labeled with "sleepy, but no difficulty remaining awake" (Åkerstedt and Gillberg, 1990). Similarly, average KSS ratings of 6 ("some signs of sleepiness") were measured after approximately 20 minutes of manual simulated driving in the study of Fors et al. (2016) in which interaction with other traffic participants was kept to a minimum. In both the above referenced studies, participants were not sleep-deprived, the study took place during daytime and the drive was designed to be monotonous, as it was the case in this study. Considering that the verbal descriptors of the average fatigue ratings in this study are comparable to the findings of Jarosch et al. (2017) and Fors et al. (2016), whose studies employed similar conditions, the results of this study are in alignment with existing findings. Also Thiffault and Bergeron (2003) found that fatigue, measured by steering wheel movements, manifest itself after 20 minutes under simulated monotonous driving conditions.

The ANOVA test result of the SSS ratings showed that there were significant differences between the different conditions, with the mean values ranging from awake ratings in the baseline drive to onset of fatigue in the tiring drive. This result is meaningful for the correlations between the SSS ratings and objective parameters. The absolute mean SSS values and the significant difference between the mean ratings show that the ratings provided by the participants cover a large range of the SSS (Figure 11). This means that the correlations reported in Chapter 4.2.3 are not restricted to only a part of the scale.

In addition to the subjective ratings, also physiological data showed significant differences between the different conditions (Table 8). The HRV (e.g. SDNN) was significantly increased at the end of the tiring drive, whereas HR was significantly decreased, which aligns with the studies of Kaida et al. (2007) and Lal and Craig (2000).

The correlation analyses of subjective fatigue ratings with objective parameters showed that even the most correlated parameters have a correlation coefficient of r=0.26 and therefore only moderate correlations were found. SDNN, blink frequency and the standard deviation of vertical gaze coordinates were positively correlated with the SSS ratings, whereas BF was negatively correlated with the SSS ratings. An increase in blink frequency during fatigue has also been found in the studies of Verwey and Zaidel (2000) and Körber et al. (2015) during simulated driving. The increase of SDNN in the tiring condition is in alignment with the study of Kaida et al. (2007) who found a correlation between the SDNN and fatigue.

The positive relation between the standard deviation of vertical gaze coordinates is in contrast with the findings of Lal and Craig (2000) who reported fast eye movements when the driver was alert, compared to limited eye movements in a fatigued state. The difference to the results of this study is due to the fact the participants gazes were focused on the distance displayed on the screen during the baseline, easy, provoking and stressful condition. Since the participants were not required to match the distance in the tiring drive, they were not required to focus on the displayed distance – resulting in larger eye movements. Therefore, this correlation should not further be considered as meaningful in the context of this study.

Studies investigating a correlation of BF with subjective fatigue that could support the results of this study were not found within the literature review. However, BF has been shown to be sensitive to workload by Pattyn et al. (2008), and in turn, low workload can provoke fatigue (see Chapter 2.1.1).

As the correlations of fatigue with the objective parameters were only moderate, these individual parameters are not suited to model fatigue. However, they serve as a screening of parameters for the fatigue model in Chapter 6, in which a combination of different signals is used to model fatigue.

The statistical analysis of the results allows for the answer (A) to RQ 1:

> A 1: Task underload in simulated highway drives induces on average after 17 minutes of driving subjective fatigue ratings that can be described with "somewhat foggy, let down" and "foggy, losing interest in remaining awake, slowed down". Physiological measures SDNN, BF and blink frequency are moderately correlated with the subjective fatigue measure.

4.4 Conclusion

The results of the driving simulator experiment proved that different driver states can be induced via traffic scenarios. Importantly, significant changes were found for perceived fatigue. These changes were not only apparent in subjective responses, but also reflected in physiological data. A foggy or slowed down feeling in participants could be induced with monotonous drives as short as 17 minutes. It is therefore concluded that a simulated monotonous highway drive is an effective method to induce passive fatigue.

5 Effect of Thermal Stimuli on Passive Fatigue While Driving

This chapter contains the description of several empirical studies that analyzed the effect of thermal stimulation. The driving studies from literature that dealt with the effect of thermal stimulation (Table 6, in Chapter 2.2.4) revealed that it is difficult to see any pattern in which conditions thermal stimulation successfully mitigates fatigue. This is because the absolute change in temperature, the duration of the temperature change, the initial temperature, the targeted body parts, the presence of sleep-deprivation, the ongoing tasks, and the dependent variables varied across these studies.

In order to gain deeper knowledge on the effect of short-term cooling on passive fatigue in a controlled vehicle setting, several simulator studies were performed. The leading RQ is:

> RQ 2: Which effects do short-term thermal stimuli have on the drivers' passive fatigue?

Different settings in terms of temperature, fanning direction and stimulus duration were tested to answer RQ 2. Chapter 5.1 elaborates on a study that tested the effect of a 6-minute stimulus directed to the upper body. In a following study, the effects of a 2-minute and a 4-minute stimulus directed to the upper body were investigated (Chapter 5.2). The third study analyzed the effect of a 4-minute stimulus directed to the legs (Chapter 5.3). In Chapter 5.4 the different results of the three preceding studies are compared and evaluated to answer RQ 2.

© Springer Fachmedien Wiesbaden GmbH, part of Springer Nature 2020
E. Schmidt, *Effects of Thermal Stimulation during Passive Driver Fatigue*,
Gestaltung hybrider Mensch-Maschine-Systeme/Designing Hybrid Societies,
https://doi.org/10.1007/978-3-658-28158-8_5

5.1 Effect of 6-Minute Upper Body Cooling[3]

The aim of this study was to investigate a decrease in fatigue caused by task underload during a monotonous drive by means of a 6-minute thermal stimulus as well as exploring the effect of cooling on the affective states of the drivers. The subordinate RQ is:

> RQ 2.1: Which effects does a 6-minute thermal stimulus of 17 °C cold air have on the drivers' subjectively perceived passive fatigue, physiological fatigue and the drivers' affective state?

In the following sections the study and results pertaining to RQ 2.1 are given. First, the applied method is explained in Chapter 5.1.1. Second, the results of subjective measures and objective measures in the different conditions and the relation between those subjective and objective measures are reported in Chapter 5.1.2. Subsequently, the results are discussed in Chapter 5.1.3 and last, the conclusions about the effectiveness of 6-minute upper body cooling are drawn in Chapter 5.1.4.

5.1.1 Method

The method section of this study is structured as follows. Chapter 5.1.1.1 gives an overview of the technical setup in which the study took place. The study design is explained in Chapter 5.1.1.2 and the sample set is described in Chapter 5.1.1.3. The dependent variables collected in the study are listed in Chapter 5.1.1.4. Last, Chapter 5.1.1.5 outlines how the collected data were analyzed.

---

[3]   This chapter is based on a previous publication: Schmidt, E., Decke, R., Rasshofer, R., & Bullinger, A. C. (2017a). Psychophysiological responses to short-term cooling during a simulated monotonous driving task. *Applied Ergonomics, 62,* 9-18. https://doi.org/10.1016/j.apergo.2017.01.017, Copyright (2017), with permission from Elsevier.

5.1.1.1 Setup

Since this study required to induce fatigue due to monotony in non-sleep-deprived subjects, a simulator study was conducted. This is because simulated driving induces fatigue faster than real driving (Chapter 3). The simulator method allowed also for reproducible traffic scenarios.

For this study, the BMW i3 apparatus was used in the curved-screen simulator (see Figure 5). Details about the simulator and the cold air generation for this setup can be found in Chapter 3.2 and Chapter 3.3. The layout of the setup is visualized in Figure 9 (p. 47). After flipping the circulation flap, it took about 60 seconds before the measured temperature at the air inlets dropped from 23 °C to 17 °C, because the air ducts of the car were warmed up to room temperature during the control period.

5.1.1.2 Study Design

The study employed a one-factor within-subject design, in which all participants drove two identical routes – one drive with short-term cooling (COOL) at the end and one drive without cooling as a control (CONT) condition (Figure 12). The order of the conditions was counterbalanced. The study began with one 5-minute familiarization drive on a highway in which participants could get used to the simulation environment. After the familiarization, the two test drives of 26 minutes followed. The duration of these drives was limited by the fact that the total length of the experiment was limited to 100 minutes because the study took place during the participants' working hours. Both test drives were highway routes with very little traffic in order to cause a monotonous task in the style of the tiring drive described in Chapter 4.1.2. Furthermore, participants were instructed to drive no faster than 120 km/h.

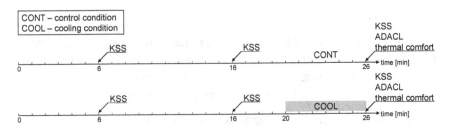

Figure 12:   Test drives with marked times of verbal assessment of fatigue and cooling
             period
Source: *Schmidt et al., 2017a*

In the COOL condition, 17 °C cold air was blown with 100 % fan intensity
in the cabin starting after 20 minutes for 6 minutes (Figure 12). A duration
of 6 minutes was chosen because of the recommendation of Van Veen
(2016) and the warning of Landström et al. (1999) which state that cooling
should last at least 3 minutes and that longer durations go at the expense
of thermal comfort. Severe cooling at 5 °C for 30 minutes has been proven
to deteriorate driving performance by Daanen et al. (2003). The reasons
for choosing upper body cooling as a stimulus are explained in Chapter 3.4.

In this study, both temperature and fan intensity were varied during the
COOL treatment to further increase the cooling due to the wind chill effect
on the exposed skin and to prevent the cold air from mixing with surround-
ing thermo-neutral air before reaching the driver.

At all other times the fans blew at an intensity of 20 % with thermo-neutral
air (23 °C). After the first monotonous test drive the participants performed
a wakening dexterity mastication task as described by Gershon et al.
(2009b) – namely shelling and eating sunflower seeds – in an attempt to
activate the participants and to create a similar initial condition before the
second monotonous drive.

5.1.1.3 Participants

Fifty BMW Group employees participated in the study in February 2016. Of those, five participants were excluded from analysis by listwise deletion due to insufficient fatigue level and nine were excluded because of technical errors. Another two participants were not able to continue the study after the familiarization drive due to simulator sickness. The final sample size was 34 participants. Study participants were 24 males and 10 females aged between 21 and 59 years (M=31.8 y, SD=11.2 y).

Details about the requirements for participation are described in Chapter 3.5. The participants were examined at different times of the day. Seven participants took part from 8 am to 10 am, eight from 10 am to 12 pm, seven from 12 pm to 2 pm, eight from 2 pm to 4 pm and four from 4 pm to 6 pm. A Kruskal-Wallis test of the initial fatigue ratings between the participants of the five test times supports that the test time did not affect the fatigue level (H(4)=1.93, p=.75).

5.1.1.4 Dependent Variables

At the beginning of the experiment, the KSS (Åkerstedt and Gillberg, 1990, translated into German by Niederl (2007)) was asked for measuring the initially perceived fatigue. In case the KSS rating stayed lower than 5 (="neither alert nor sleepy") for a participant throughout the experiment the datasets were excluded from the statistical analysis because the study requirement of passive driver fatigue was not met.

The KSS was also asked verbally during the experiment via a microphone-speaker system after the 6th and 16th minute of the drives without stopping the drive (Figure 12). Schmidt et al. (2011) showed that the driver's state can be verbally assessed every 5 minutes without resulting in a continued activation of the driver. Verbal assessment only has a short-term effect on the subjects' vigilance level and the physiological measures are back to pre-communication levels after a maximum of 2 minutes.

After each drive, the participants responded one more time to the KSS. Furthermore, the ADACL following Thayer (1989) (German version by Imhof (1998)) was asked to measure the drivers' affective states in terms of their perceived positive or negative activation, and the thermal sensation scale described in EN ISO 14505-3 (2006) as well as the Bedford scale (Bedford, 1936) for rating thermal comfort.

The post-questionnaire included yes-no questions about the liking and the perceived effectiveness of the cooling against passive fatigue, which were adopted from Van Veen (2016). Van Veen (2016) asked the questions "Did you like the cooling?" and "Do you think that the cooling helps against fatigue?" which were translated into German by the author of this thesis to the phrases "War die Kühlung angenehm?" and "Denken Sie, dass die kurzzeitige Kühlung die Müdigkeit reduziert hat?" (see A2.1, p. 179). In the comment fields of the questionnaire, participants could also describe their impressions of the cooling or other thoughts. After the participants had experienced both drives, they were also asked to indicate which of the two drives they preferred.

Besides the subjective data, also physiological signals were recorded. Details about the recording of the physiological data are described in Chapter 3.6.

5.1.1.5 Data Analysis

For details about the data analysis, please see Chapter 3.7. The majority of data were not normally distributed. Therefore, two-sided Wilcoxon tests were used for the statistical analyses of differences of means. Only the responses of participants to the thermal sensation according to EN ISO 14505-3 (2006) were normally distributed and analyzed using t-tests.

5.1.2 Results

The results on the subjective assessments of fatigue and activation are presented in Chapter 5.1.2.1. Chapter 5.1.2.2 reports on the results of the physiological measures. Last, Chapter 5.1.2.3 lists the results of correlation analyses between subjective and objective data.

5.1.2.1 Subjective Data

The results of the KSS are shown in Figure 13. There were no statistically significant differences between the fatigue levels after the 6th and 16th minute of the drives COOL and CONT. The fatigue increase from minute 6 to 16 was significant in both conditions (Z=-4.46, p<.001 for COOL and Z=-4.92, p<.001 for CONT), and so was the increase from minute 16 to 26 in the CONT drive (Z=-3.90, p<.001). As hypothesized, the average fatigue level after 26 minutes was significantly different between the two conditions (Z=3.64, p<.001). In the COOL drive, the fatigue level after 26 minutes was not different (Z=0.88, p=.38) from the level after 16 minutes of driving.

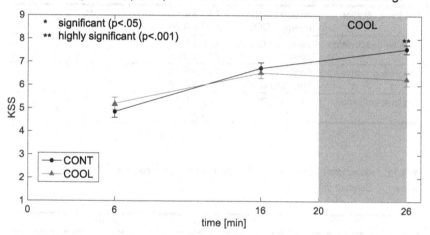

Figure 13: KSS ratings (M and SE) over the drives CONT and COOL
Source: *Schmidt et al., 2017a*

The results of the ADACL showed that the construct "energy" in the COOL drive increased compared to the CONT drive (Z=-3.35, p<.001) (Table 10). The construct "tiredness" decreased significantly after the short-term cooling (Z=3.40, p<.001). There were no significant differences in "tension" (Z=-1.05, p=.30) and "calmness" (Z=0.66, p=.51).

Table 11 shows the results of the questionnaires regarding the impressions of the cooling. A percentage of 59 % of the participants liked the cooling and the comments of these people often included that the cooling was perceived as refreshing and that it had a wakening effect.

Table 10:   ADACL results (M and SD) and significance test results between means
            (- not significant, ** significant (p<.001))

Source: Schmidt et al., 2017a

Psychological construct		M, SD (based on 1-4 scales)	Wilcoxon test result
positive activation	energy CONT	1.51 ± 0.57	**
	energy COOL	1.95 ± 0.78	
positive deactivation	calmness CONT	3.14 ± 0.41	-
	calmness COOL	3.06 ± 0.61	
negative deactivation	tiredness CONT	3.25 ± 0.59	**
	tiredness COOL	2.77 ± 0.76	
negative activation	tension CONT	1.42 ± 0.41	-
	tension COOL	1.49 ± 0.56	

Table 11:   Questionnaire answers to the impressions of the cooling
Source: Schmidt et al., 2017a

	Yes [%]	No [%]
Did you like the cooling?	59	41
Do you think that the cooling helps against fatigue?	88	12
	With cooling [%]	Without cooling [%]
Would you prefer a monotonous drive with or without short-term cooling?	78	22

The comments of the people disliking the cooling revealed that the cooling felt too cold in their faces. This aligns with an average cool thermal sensation of -1.5 (-1 meaning "slightly cool", -2 meaning "cool" according to EN ISO 14505-3 (2006)) compared to the neutral thermal sensation of 0.2 in the control condition. The majority of people (88%) thought that cooling reduced their fatigue (Table 11). The comments here included hints that the wakening effect of the cooling is only short-lived and participants indicated that they would have felt tired again if the drive had continued for a longer duration because they got used to the stimulus. After completion of both drives, 78% of the drivers would prefer a monotonous car drive with short-term cooling (Table 11).

5.1.2.2 Physiological Parameters

Figure 14 shows the graphs of the HR for both drives (averaged over all participants). The vertical lines at minute 6 and 16 in the graph indicate where fatigue was verbally assessed. HR decreased with cooling onset and stayed significantly lower than in the CONT drive during the first 3 minutes of cooling.

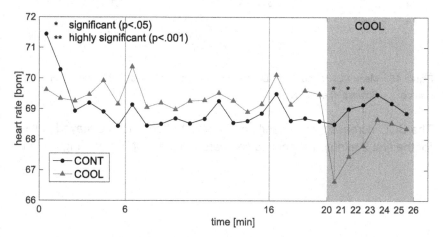

Figure 14:   Mean heart rate over the course of the drives CONT and COOL
Source: *Schmidt et al., 2017a*

The two time domain HRV measures, SDNN and RMSSD, were increasing over the course of both drives which is an indicator for increasing fatigue (Figure 15). There were no significant differences between the two conditions at any point of the drive. Neither did the cooling affect BF nor gaze variability.

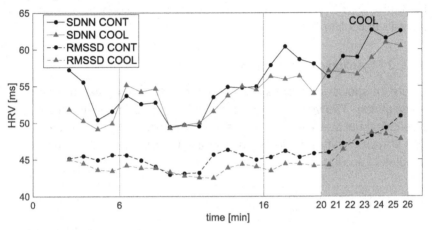

Figure 15:   Mean SDNN and RMSSD over the course of the drives CONT and COOL
Source: *Schmidt et al., 2017a*

The SCL increased significantly at the onset of cooling and stayed higher for the first 2 minutes of cooling compared to the CONT drive (Figure 16).

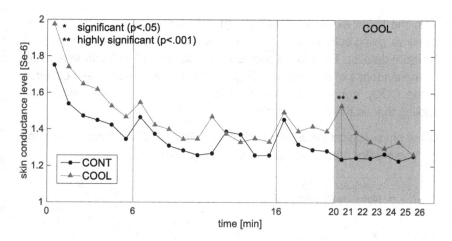

Figure 16: Mean skin conductance over the course of the drives CONT and COOL
Source: *Schmidt et al., 2017a*

The pupil diameter significantly increased during the entire duration of the cooling period compared to the CONT condition. Figure 17 shows that the largest increase of the diameter during cooling, was measured during the beginning of the cooling period.

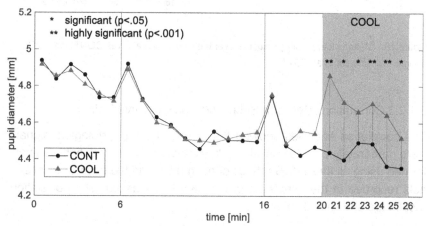

Figure 17: Mean pupil diameter over the course of the drives CONT and COOL
Source: *Schmidt et al., 2017a*

With the facial expression recognition software evidence values for eye closures, using action unit number 43, could be extracted (see Chapter 3.6 for more details). The results showed that cooling had an immediate effect on the evidence of eye closures (Figure 18). The evidence of eye closures in the COOL condition was significantly lower for the first 3 minutes of cooling than in the CONT condition.

Other physiological measures such as LF, HF and BF were not affected by the cooling (for details see Schmidt et al., 2017a).

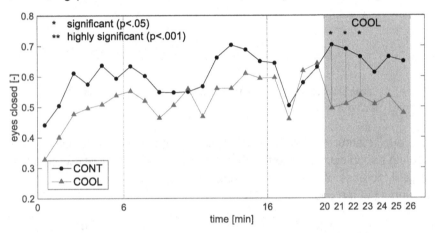

Figure 18:   Mean evidence of eye closure over the course of the drives CONT and COOL
Source: *Schmidt et al., 2017a*

5.1.2.3 Correlations Between Subjective and Objective Data

The correlation analyses showed that a few of the physiological signals were moderately correlated with the KSS ratings. The correlation was performed using all three KSS ratings of each drive and the physiological signals recorded in the minute before the KSS was asked. The correlation results are shown in Table 12.

Table 12: Correlations of KSS ratings with physiological data
Source: Own table

Signal	Correlation with KSS
SDNN	r=0.24, p<.001
RMSSD	r=0.31, p<.001
LF	r=0.20, p<.05
change in pupil diameter (compared to first driving minute)	r=-0.29, p<.001

All other parameters that were extracted from the physiological data in this study (see Chapter 3.7) did not yield significant correlations or had a correlation coefficient under $|r|=0.2$.

5.1.3 Discussion

The goal of this study was to investigate the psychophysiological effect of short-term cooling when the driver is fatigued due to monotony. Previous studies in literature (see Chapter 2.2.4) used similar in-car treatments (Landström et al., 1999; Landström et al., 2002; Reyner and Horne, 1998), but only a few studies investigated the effect of cooling on non-sleep-deprived subjects (Schwarz et al., 2012; Van Veen, 2016). In those studies, subjectively rated sleepiness was not significantly decreased. Therefore, a study was conducted with a larger temperature and fan intensity difference, and a different fanning direction.

Fatigue induction

One requirement of the study was the induction of fatigue in the participants by means of monotonous driving. As the analysis of the KSS ratings (Figure 13) showed, participants perceived increasing fatigue throughout the drive. This was supported by the increase of HRV, measured with SDNN and RMSSD (Figure 15). The verbal assessment of fatigue is believed to have activated the drivers physiologically because HR (Figure 14), SCL (Figure 16) and pupil diameter (Figure 17) increased after minute 6 and 16, when participants were required to respond to the KSS. The activation lasted only shortly and the parameters returned to the pre-as-sessment levels after 1 minute, which aligns with the observations of Schmidt et al. (2011).

Cooling effect from subjective measures

The evaluation of the questionnaires measuring fatigue and the activation of the driver showed that the cooling decreased perceived fatigue (Fig-ure 13) and subjects felt significantly more energetic (Table 10). The ma-jority of the participants believed that cooling reduced fatigue (Table 11). However, a non-negligible percentage of the drivers (41 %) did not like the cooling because it felt too cold (Table 11). Similar negative-feedback rates (53 %) were yielded by hand cooling in the study of Van Veen (2016).

Cooling effect from objective measures

The enhanced subjective perceptions on alertness could also be confirmed from several physiological responses of the drivers. An SNS activation was indicated by the increase of SCL and pupil diameter (Figure 16 and Fig-ure 17) during the cooling period. These reactions are symptoms of auto-nomic activation due to sensory stimulation (Bradley et al., 2008; Goldwa-ter, 1972). The two peaks in minute 6 and 16 in both graphs can be at-tributed to the verbal KSS ratings provided at these times, which caused these task-related pupillary responses (Beatty, 1982).

The reduced evidence of eye closures during the first 3 minutes of cooling (Figure 18) was another observation supporting the reduction of fatigue through cooling. On the one hand, the reduction of eye closure evidence during the thermal stimulus with increased airflow is surprising because the study of Acosta et al. (1999) found an increase in blink rates when eyes were exposed to an airflow. On the other hand, blink rates and PERCLOS have been shown to be lower in alert states (Jarosch et al., 2017; Körber et al., 2015). In this study, the awakening effect of thermal stimulation on eye closures must have been therefore stronger than the effect of airflow on eye blinks.

The reduction of HR (Figure 14) during cooling does not seem to align with the increased wakefulness of the driver at first glance. As the correlation analyses in Chapter 5.1.2.3 have shown, however, HR did not prove to be a good indicator for passive fatigue during simulated driving. The reduction of HR was probably a physiological response of the short-term cooling, which has been observed with facial fanning in non-vehicle settings (Collins et al., 1996; Hayward et al., 1976; LeBlanc et al., 1975, 1976, 1978). Researchers have suggested that the decrease of HR is caused by a vagal reflex in which facial receptors initiate a stimulation of the trigeminal nerve (Collins et al., 1996; LeBlanc et al., 1976). As the study of LeBlanc et al. (1976) has shown, the reflex occurs with fanning at temperatures as high as 25 °C. Both increasing wind speeds and decreasing temperatures were found to strengthen the bradycardia.

The increase in HR through hand cooling observed by Van Veen (2016) in a simulator study (Table 6) is not contradictory to this study since it has been proven that hand and facial cooling cause different HR responses (LeBlanc et al., 1975, 1978).

Against expectations, the variability in gaze direction, time and frequency domain HRV measures and BF did not change through the cooling (for details see Schmidt et al., 2017a). This differed from the results of Lal and Craig (2000) who found that eye movements were fast in a state of wakefulness compared to little eye movements during fatigue. The lack of effects on HRV or BF was also in contrast to the previous analyses (Chapter 4.2.2) in which those variables were affected by the treatment. In the past study (Chapter 4), however, changing traffic scenarios affected these parameters instead of in-vehicle treatments like turning on the AC. It is assumed that the effect of cooling was not strong enough to be reflected in gaze variability, HRV and BF.

Correlation between subjective an objective fatigue measures

The correlation analyses of subjective fatigue ratings with objective parameters showed that even the most correlated parameters have a correlation coefficient of under $|r|=0.35$ and therefore only moderate correlations were found (Chapter 5.1.2.3, Table 12).

It is also noticeable that in the two studies (fatigue induction study of Chapter 4 and cooling study of Chapter 5.1), different signals were the best indicators for fatigue as shown in Table 9 (p. 63) and Table 12 (p. 79). First, this could be due to the different subjective scales, which was the SSS in the first, and the KSS in the second study. Another reason could be that the drivers' states were induced in different ways. In the first study, different driver states were induced by means of traffic scenarios that varied in traffic density and following task. In the second study, only the time-on-task and the cooling effect influenced the fatigue ratings.

The various traffic scenarios imposing different workloads in the first study caused larger variations in BF, gaze coordinates and blink frequency which were evaluated at the end of each drive (Chapter 4.2.3). This could be the reason these signals correlated moderately with the SSS ratings. The second study caused larger variations in HRV over the course of the drive, which was also longer (26 minutes) than the tiring drive of the first study (17 minutes). Therefore, the HRV measures correlated moderately with the KSS.

Since the correlation of fatigue with the objective parameters was only moderate, these single parameters are not suited to model fatigue. However, they serve as a screening of parameters and it can be concluded, that in the studies of this thesis the physiological parameters, especially HRV, proves to be a good indicator. Because the single parameters did not yield high correlations with fatigue, Chapter 6 will analyze whether combinations of different signals reveal patterns that are better indicators of fatigue.

Taking all study results into account, RQ 2.1 can be answered as follows:

> A 2.1: A 6-minute thermal stimulus of 17 °C air significantly reduces drivers' subjectively perceived fatigue. The stimulus also reduces physiological fatigue as a significant increase in skin conductance and pupil diameter was measured which indicates an SNS activation. Moreover, the evidence in eye closures was significantly reduced. In terms of the affective state, the drivers felt more energetic.

Habituation

The increase of SCL during cooling was only significant for the first 2 minutes of cooling (Figure 16). After that, there was no significant difference between the SCL measured during the COOL and CONT condition. This was most likely because participants got used to the thermal stimulation after some time. This assumption could partly be substantiated by the text written in the comment fields by the participants.

Limitations

Limitations of the study were that both air temperature and air movement were varied in the COOL condition and the stronger fanning came with an increase in background noise. Therefore, it is possible that the measured data were not only affected by a colder temperature per se. The observed effects were probably due to a combined effect of thermal, auditory and tactual stimulation because of cold air, fan noise and the air pressure on the skin. The fanning caused a constant noise for the 6 minutes of cooling on top of the driving noise. The change in noise level influenced very likely the first seconds of the physiological measurements due to an orienting response. An orienting response is a phenomenon, which describes a set of physiological reactions to an unexpected stimulus of auditory, visual or tactual nature (Bradley, 2009; Frith and Allen, 1983; Graham and Clifton, 1966). As the studies of Bradley (2009) have shown, the orienting response lasts less than a few seconds and causes HR deceleration and an increase in SCL. This means that the reported reactions of these data in this study were probably affected by this phenomenon because of the increased noise level and tactual stimulation through airflow in the COOL condition.

A pre-study in which 6 minutes of fanning was tested without a change in temperature on a sample of six people, can qualitatively substantiate this hypothesis on the secondary effect through the orienting response caused by the auditory and tactual disturbance. In this pilot study, the fanning intensity was changed from 20 % to 100 % with a constant temperature of 23 °C. In this case, the physiological responses were as follows: SCL and pupil diameter increased noticeably in the first minute of fanning, but returned to pre-fanning levels after 1 minute. HR decelerated with onset of fanning and returned to pre-fanning level after 2 minutes. The HR reaction could not solely be attributed to the orienting reflex, which only lasts for seconds. Instead, this reaction was the result of the vagal reflex explained earlier.

Taking the results of the pilot test with constant temperature and results from the literature into account, it could be concluded that the secondary effect of changes in fan intensity and noise level affected the first minute of the electrodermal and pupillary responses. The HR measured in this study was affected for the entire COOL period by the fanning because both increasing wind speed and decreasing temperature have shown to decrease HR over the course of several minutes (LeBlanc et al., 1976). A combined fanning and cooling lowers the HR even more (LeBlanc et al., 1976). Since in this study, a continued effect on SCL and especially on pupil diameter during the COOL treatment was measured, it is assumed that this continued activation is due to the lower air temperature.

5.1.4 Conclusion

In this study both objective and subjective responses indicated that thermal stimulation caused psychophysiological arousal and decreased fatigue. According to participants' KSS ratings and self-evaluation via the ADACL, the drivers felt more awake and activated after the cooling. The pupillary and electrodermal response indicated an SNS activation which suggests a state of alertness in the driver. Furthermore, the amount of eye closures was decreased through the cooling, which is also a sign of decreased fatigue.

The COOL treatment consisted of a 6-minute temperature reduction to 17 °C and an increased airflow accompanied by a constant audible noise. The driver was therefore exposed to a multi-sensory stimulation: thermal, tactile and auditory. Therefore, all measured subjective and objective differences between the conditions could not be attributed solely to the cooling stimulus.

Based on the length of the effects on HR, SCL, pupil diameter and eye closure evidence, a shorter cooling duration of 2 to 4 minutes is recommended. After this period, continued cooling will not activate the driver any further, as the objective data evaluation suggests. This would also align with the participants' impression that they got used to the cooling after a few minutes and that they only felt more awake in the short term. These impressions are also supported by the results of Reyner and Horne (1998) who found that a cooling period of 2 hours at an even lower temperature did not affect sleepiness, whereas in the first minutes, significantly lower KSS ratings were recorded. Following this thought, a duration of cooling longer than a few minutes, will worsen the driving experience since it goes at the expense of the thermal comfort and deteriorates driving performance (Daanen et al., 2003). Further research should focus on identifying the minimum temperature drop for which an activation of the driver can be measured.

The study results showed that short-term cooling at 17 °C was a countermeasure against passive fatigue, while driving in monotonous road conditions, because it caused physiologically measureable activation. It could also be concluded from the collected data that cooling only had a short-lasting effect on fatigue. To highlight the effect of the cooling on the perceived fatigue it should be pointed out that the fatigue level after minute 26 in the COOL condition was comparable to the participants' fatigue level after 16 minutes of driving (Figure 13).

5.2 Effect of 2-Minute and 4-Minute Upper Body Cooling[4]

The first study (Chapter 5.1) showed that the activation of the driver was only short-lived and that extended cooling durations would not continue to reduce fatigue because of habituation effects. This was observed from the time-limited effects in SCL and blink occurrence which only lasted about 2 to 3 minutes, as well as from the comments of the drivers about the effectiveness of cooling (Chapter 5.1.2.3). According to these observations, shortening the cooling duration might be preferable to increase driver comfort while still maintaining effectiveness. From these insights, a new simulator study was conceived that investigated different cooling durations that should be both long enough to invoke SNS effects, but differ in the expected likelihood of habituation.

To investigate the decrease in subjective and physiological fatigue during a monotonous drive by means of a 2-minute and a 4-minute thermal stimulus, a simulator study was conducted. The RQ this study adresses is:

> RQ 2.2: Which effects do a 2-minute and a 4-minute thermal stimulus of 15 °C cold air have on the drivers' subjectively perceived fatigue and physiological fatigue?

The study to answer RQ 2.2 is described in the following sections. First, the study method is presented in Chapter 5.2.1, followed by the report of the results in Chapter 5.2.2. In Chapter 5.2.3 the results are discussed and, last, conclusions about the effect of the different stimulus durations are drawn in Chapter 5.2.4.

---

[4]   This chapter is based on a previous publication: Schmidt, E., & Bullinger, A. C. (2019). Mitigating passive fatigue during monotonous drives with thermal stimuli: Insights into the effect of different stimulation durations. *Accident Analysis & Prevention, 126,* 115-121. https://doi.org/10.1016/j.aap.2017.12.005, Copyright (2019), with permission from Elsevier.

5.2.1 Method

The description of the method is divided in five Chapters. Chapter 5.2.1.1 refers to the technical setup in which the study took place. The study design is explained in Chapter 5.2.1.2, followed by the sample set description in Chapter 5.2.1.3. The dependent variables collected in the study are listed in Chapter 5.2.1.4. Last, Chapter 5.2.1.5 gives an overview of the analyses performed on the collected data.

5.2.1.1 Setup

For this study, the BMW i3 apparatus was used in the flat-screen simulator (see Figure 7). The cold air was generated with the car's AC and could be controlled remotely at the investigator's desk (Chapter 3.2 and 3.3).

5.2.1.2 Study Design

The aim of this study was to gain further insights into the effects of different cooling durations to reduce passive fatigue. Passive fatigue of non-sleep-deprived participants was induced by means of a simulated monotonous driving task and thermal stimuli of different durations were applied as fatigue countermeasures.

The study began with one 5-minute familiarization drive on a highway in which participants could get used to the simulation environment. All drives were highway routes in the style of the tiring drive in Chapter 4.1.2.

The study employed a repeated-measures design with the cooling duration as the independent variable. The factor was tested at three levels: control (CONT – no cooling), 2 minutes of upper body cooling (COOL2) and 4 minutes of upper body cooling (COOL4). The cooling durations were chosen based on the hypothesis that a stimulus has to be equal to or longer than 2 minutes and shorter than 6 minutes. These duration limits were derived, in part, from the review of Van Veen et al. (2014) who suggested that cold stimuli (e.g. 5 °C) need to be applied for at least 2 minutes to elicit an SNS activation. A 6-minute stimulus was investigated in Chapter 5.1 and it was found, that the stimulation duration should be shorter. Taking these considerations into account, 2- and 4-minute stimulus durations were chosen because these are equidistant in the given interval. At all other times a thermo-neutral climate at 24 °C was maintained. The reasons for choosing upper body cooling as a stimulus are explained in Chapter 3.4.

To induce passive fatigue, participants drove on monotonous roads for 15 minutes (Figure 19). Based on the observations from the pre-study on inducing passive fatigue by means of traffic scenarios (Chapter 4), it is concluded that this period of monotonous driving in a simulator evokes sufficiently high fatigue ratings. After the fatigue induction period, one of the two thermal stimuli – COOL2 or COOL4 – was applied in a counterbalanced way. The stimulus was followed by a period of 6 minutes, in which participants kept driving without thermal stimulation. In this period, subjective and physiological fatigue were expected to return to pre-cooling levels. The reason for this shorter period before the second stimulus resulted from the first cooling study (Chapter 5.1) that showed that thermal stimulation of the upper body increased physiological arousal and subjective alertness, but drivers were still more fatigued after the stimulation compared to the start of the drive. Therefore, it was concluded, that it is not necessary to have the participants drive for another 15 minutes, as they did, before the first stimulus. After a period of 6 minutes, the respective alternative thermal stimulus was then applied, followed by another monotonous period of 6 minutes (Figure 19). The total length of the drive was 33 minutes and is referred to as COOL drive.

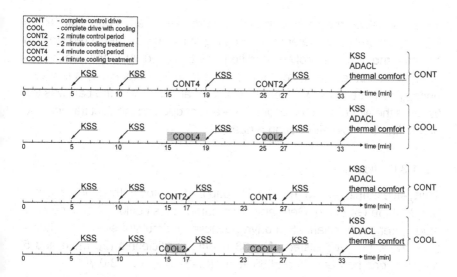

Figure 19: Test drives with marked times of verbal assessment of fatigue and cooling durations

Source: *Schmidt and Bullinger, 2019*

In order to generate corresponding control settings, the participants also completed another 33-minute long drive with no stimuli (CONT). The order in which the participants completed the two drives was counterbalanced. After the first drive the participants were tasked to shell and eat sunflower seeds for approximately 5 minutes – a dexterity mastication task which, according to the research of Gershon et al. (2009b) activates drivers. This wakening effect was used between the two drives to create a similar initial condition before the second monotonous drive.

Air at a temperature of 15 °C was used for the cooling condition. This temperature was cooler compared to the stimulus of 17 °C applied in the previous study in Chapter 5.1. However, the 17 °C stimulus of Chapter 5.1 was combined with an increased airflow. In order to have a noticeable reduction in climate without the use of additional wind chill effects in this study, the lower temperature of 15 °C was chosen. The fan intensity stayed at 30 % at all times. After setting the temperature to 15 °C, it took about 60 seconds before the measured temperature at the air vents dropped from 24 °C to

15 °C. However, the change in temperature was clearly noticeable within the first seconds and therefore the cooling duration was measured starting from the moment the cooling condition was triggered. After the cooling period, the rise of temperature from 15 °C back to 24 °C took only a few seconds. This is because warmer air, stored in an air duct of the car, was led towards the air vents of the cabin by means of opening the duct flap instead of sending the cold air through a heater.

5.2.1.3 Participants

Thirty-five employees of the BMW Group participated in the study in October 2016, in Garching, Germany. Two participants could not continue the study after the familiarization drive because of simulator sickness. Therefore, the sample consisted of n=33 healthy participants (28 male and 5 female) aged between 20 and 66 years (M=31.9 y, SD=13.0 y).

Details about the requirements for participation are described in Chapter 3.5. The average sleep duration in the night before the study session was M=6.8 h, SD=0.9 h. All tests took place from 8 am to 5 pm. After the 5-minute familiarization drive the average KSS rating of the participants was M=4.3, SD=1.5, which lies between "alert" and "neither alert nor sleepy".

5.2.1.4 Dependent Variables

The KSS (Åkerstedt and Gillberg, 1990, translated into German by Niederl (2007)) was asked after the familiarization drive as well as verbally via a microphone-speaker system after the 5th and 10th minute of the drives without stopping the drive, as visualized in Figure 19. Furthermore, the KSS was asked directly after the end of the COOL2 and COOL4 stimuli and at the corresponding points in time in the CONT drive (Figure 19). To identify the sections in the CONT drive which corresponded to one of the cooling conditions, the abbreviations CONT2 and CONT4 were chosen and are marked in Figure 19.

After each of the drives, the participants filled out several questionnaires including the KSS, the ADACL (Thayer, 1989; German version by Imhof, 1998), which was surveyed to measure the drivers' positive or negative activation, as well as the ASHRAE scale (American Society of Heating, Refrigerating and Air-Conditioning Engineers, ASHRAE, 1966) and the Bedford scale (Bedford, 1936) for evaluating thermal sensation and thermal comfort. In case of the COOL drive, participants were asked to rate the thermal comfort they perceived during the coolest part of the drive.

In the case of the COOL drive additional questions were asked to evaluate the impressions of the drivers after completion of the KSS, ADACL, ASHRAE and the Bedford scale. The central questions were the same as in the study of Van Veen (2016), with the goal of obtaining whether the drivers liked the cooling and whether they perceived that the cooling reduced fatigue (see A2.2, p. 184). The participants could also describe their thoughts on the thermal stimulus and its effectiveness against passive fatigue in comment fields. After the participants completed both drives, they were asked which of the drives, CONT or COOL, they preferred.

Details about the recording of the physiological data are described in Chapter 3.6.

5.2.1.5 Data Analysis

The physiological data filtering and analysis are described in Chapter 3.7. The HR and pupil diameter data were normally distributed and it was hypothesized that HR would decrease and pupil diameter would increase with cooling, as previous study results have shown (Chapter 5.1). Therefore, one-sided t-tests were used for statistical analyses of the cooling effect. For all other statistical testing, e.g., the time-on-task effect or testing for differences between conditions at times without cooling, two-sided tests were used.

The evaluation of the results of the normally distributed KSS ratings and the not normally distributed ADACL was done using one-sided t-tests and Wilcoxon tests, respectively, as it was hypothesized that the constructs "energy" and "tension" increase whereas "tiredness", "calmness" and the KSS ratings decrease in the COOL condition.

5.2.2 Results

In the following, the results on the time-on-task effect are presented in Chapter 5.2.2.1. Chapter 5.2.2.2 contains the results in terms of the effect of the thermal stimuli of different durations.

5.2.2.1 Time-on-Task Effect

Participants perceived increasing fatigue levels over the course of the drive. Figure 20 shows the average KSS ratings of all participants after minute 5, 10 and 33 for both CONT and COOL drive.

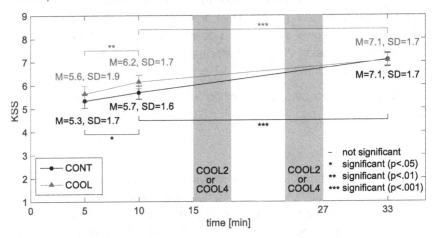

Figure 20:   KSS ratings (M and SE) over the course of the drives CONT and COOL
Source: Schmidt and Bullinger, 2019

There were no significant differences for the KSS ratings between the two conditions at any of the times, as two-sided t-tests showed (t(32)=-1.1, p=.30 at 5 minutes; t(32)=-1.6, p=.12 at 10 minutes; t(32)=0.1, p=.93 at 33 minutes). The insignificant difference of KSS ratings at the end of the drives indicated that the second cooling did not result in a lasting effect. This supports the hypothesis that 6 minutes after the stimulus, fatigue continues. Subjective fatigue increased significantly from minute 5 to 10 in both drives (CONT: t(32)=-2.7, p=.01; COOL: t(32)=-3.2, p=.004) and from minute 10 to 33 (CONT: t(32)=-4.2, p<.001; COOL: t(32)=-4.6, p<.001), as confirmed by the two-sided t-tests.

5.2.2.2 Effect of Stimulus Durations

The subjective assessments indicated that drivers felt more alert after the COOL2 and COOL4 treatments. A comparison of the KSS ratings given after COOL2 (M=6.2, SD=2.1) and at the respective time of the control drive, CONT2 (M=6.7, SD=1.4), showed that KSS ratings were significantly lower after the COOL2 condition (t(32)=-1.9, p=.03) (Figure 21a). An even larger difference (t(32)=-2.2, p=.02) was found between the KSS ratings given after COOL4 (M=6.1, SD=1.9) and those after the CONT4 assessments (M=6.7, SD=1.6) (Figure 21b). When comparing the KSS ratings of the COOL2 and the COOL4 stimuli with each other, no significant difference between these two was found (t(32)=0.5, p=.64).

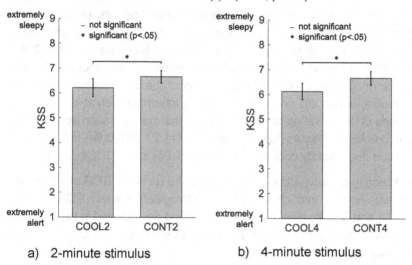

a) 2-minute stimulus   b) 4-minute stimulus

Figure 21: KSS ratings (M and SE) given directly after a) COOL2, CONT2 and b) COOL4, CONT4

Source: Schmidt and Bullinger, 2019

As explained earlier, the participants were tasked to perform a dexterity mastication task between the drives, which was expected to reduce the passive fatigue that had developed over the course of the first drive. A comparison of the KSS ratings provided after 5 minutes of driving in the first (M=5.2, SD=1.9) and second (M=5.8, SD=1.7) drive, showed that the difference in the ratings was non-significant (t(32)=-1.4, p=.18). In addition, the differences of the ratings provided after 10 minutes (t(32)=-1.2, p=.25) and 33 minutes (t(32)=0.1, p=.95) were non-significant between the first and the second drive.

The evaluation of the results of the ADACL showed that none of the hypotheses in Chapter 5.2.1.5 could be confirmed, as all changes were on a non-significant level. However, there was a trend for increased "energy" (Z=-1.35, p=.09) and decreased "tiredness" (Z=1.34, p=.09) after the COOL drive.

An evaluation of the questions regarding the participants' impressions of the cooling revealed that 85 % of the drivers liked the cooling, and 100 % believed that cooling reduced fatigue. After experiencing both the CONT and COOL drive, 91 % of the drivers preferred a monotonous drive with short-term cooling. This means that some of the people who did not like the cooling would still prefer to drive with intermittent short-term cooling because of its activating effect. The mean thermal sensation according to the ASHRAE Scale was "slightly warm" (M=1.27, SD=0.84) in the CONT condition and "slightly cool" (M=-1.03, SD=1.05) in the COOL condition.

In addition to the subjective assessments, the recorded ECG and eye data provided insights into changes in physiological arousal. ECG data of n_{ECG}=31 participants was of usable quality and are included in the analysis. During the COOL2 stimulus, the HR was lower than in the CONT2 condition (Figure 22).

In the first cooling minute this difference was a trend (t(30)=1.6, p<.059), but only in the second minute was HR significantly different (t(30)=1.8, p=.04) between the conditions.

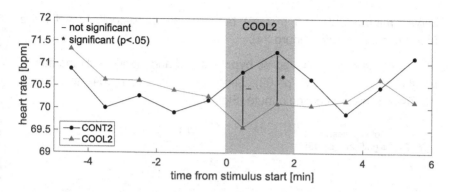

Figure 22:   Mean heart rate before, during and after COOL2 as well as CONT2
Source: *Schmidt and Bullinger, 2019*

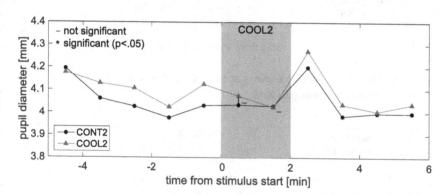

Figure 23:   Mean pupil diameter before, during and after COOL2 as well as CONT2
Source: *Schmidt and Bullinger, 2019*

Pupil tracking data was successfully retrieved from $n_{pupil}=30$ participants. There were no differences in pupil diameter (Figure 23) between the CONT2 and COOL2 condition at any point in time. The peaks of the pupil diameters after the 2-minute cooling stimulus were due to the verbal KSS rating given at this time (Figure 23).

Similar results could be found for the COOL4 stimulus (Figure 24). As in the COOL2 evaluation, the HR decrease became significant for the second minute of cooling ($t(30)=1.8$, $p=.04$). It decreased even further in the third cooling minute (t(30)=2.8, p=.005), but returned to pre-cooling level in the fourth cooling minute (Figure 24).

The pupil diameter behaved as hypothesized and significantly increased for the first (t(29)=-2.3, p=.02), second (t(29)=-1.8, p=.04) and fourth (t(29)=-2.3, p=.01) cooling minute (Figure 25).

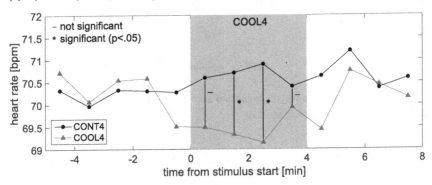

Figure 24:   Mean heart rate before, during and after COOL4 as well as CONT4
Source: Schmidt and Bullinger, 2019

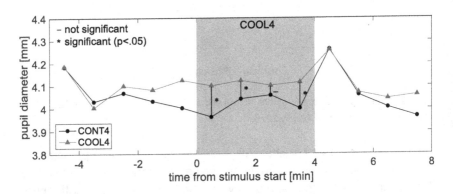

Figure 25:   Mean pupil diameter before, during and after COOL4 as well as CONT4
Source: Schmidt and Bullinger, 2019

SCL did not change significantly between the conditions COOL4 and CONT4 or COOL2 and CONT2.

5.2.3 Discussion

This study investigated different stimulus durations with respect to their effect on reducing passive fatigue during monotonous driving. Based on the results of the KSS rating given after the COOL2 and the COOL4 condition, compared to the rating after the CONT2 and CONT4 condition, it could be concluded that both stimuli reduced subjective fatigue (Figure 21). The COOL2 stimulus significantly decreased KSS ratings, whereas the physiological measures were less affected. The absolute mean difference of the KSS ratings between COOL2 and CONT2, however, was relatively low: 0.5 on a scale from 1 to 9. Similarly, the COOL4 stimulus caused reduced subjective fatigue as well as increased physiological arousal. In terms of the subjective ratings, the KSS ratings significantly decreased. The absolute mean difference of 0.6 on the KSS scale was also low (Figure 21b).

The results of the subjective ratings in this study aligned with similar studies (Landström et al., 1999, 2002), which used facial or head cooling that lasted no longer than 8 minutes. In these studies, the KSS ratings were also significantly lower after the cooling treatment. The results of this study are in contrast to the findings on subjective ratings from Van Veen (2016) and Reyner and Horne (1998) who did not see an effect on subjective ratings. The study of Van Veen (2016) yielded non-significant differences in subjective ratings, which may be due to the fact that only the hands were cooled. The non-significant changes in subjective ratings produced by the 2-hour stimulus of Reyner and Horne (1998) were likely due to habituation effects.

The observed increase in pupil diameter during the COOL4 stimulus (Figure 25) was very likely a response of an SNS activation and can be interpreted as increased physiological arousal. As seen in the study of Chapter 5.1, this supports the drivers' alertness increasing in the short term. In addition to pupil dilatation, the thermal stimulus caused a decrease in HR (Figure 24), which was triggered by the trigeminal nerve stimulation through facial receptors (Collins et al., 1996; LeBlanc et al., 1976).

The reduction in HR due to facial cooling in this study was in contrast to the observations on hand cooling from Van Veen (2016) who recoreded an acceleration of HR. It has been shown, though, that hand and face cooling evoke different reactions in HR (LeBlanc et al., 1975). The HR decrease in this study was not significantly lower in the first minute of cooling, but only for the second (and third in COOL4) cooling minute (Figure 22 and Figure 24). This was contrary to the first cooling study in Chapter 5.1, in which an immediate decrease was measured. The reason for this could be related to the fan intensity. In this study, the fan intensity was fixed and it took about 60 seconds before the temperature at the air vents dropped from 24 °C to 15 °C, therefore, the stimulus was not reflected in the HR response in the first minute.

Comparing both stimulus durations, the absolute decrease in subjective fatigue was larger for the COOL4 condition. In addition to this result of subjective evaluation, the analysis of physiological data supported a larger effect of the COOL4 stimulus, as can be seen by the continued effects in HR and pupil diameters measured during the COOL4 stimulus event. Thus, the COOL4 stimulus proved to be more effective than the COOL2 stimulus. As studies with longer thermal stimuli have shown (2 hours by Reyner and Horne (1998) and 6 minutes in Chapter 5.1), these longer-lasting stimuli are more subject to habituation. Therefore, it is concluded that the 4-minute stimulus is better suited to reduce passive fatigue and is still less subject to habituation.

In contrast to the KSS ratings provided directly after the stimulus, the ADACL failed to show a continued activation of the driver. This questionnaire was asked 6 minutes after the second cooling stimulus ended. This delay between the stimulus and the ADACL questionnaire likely caused the non-significant decrease of the construct "tiredness" in the ADACL, because the cooling did not have a continued wakening effect.

The responses of the drivers concerning their feelings towards the cooling and their preference on how to conduct a monotonous drive, showed that the majority of people liked the cooling (85 %) and that most participants would prefer to drive in a monotonous setting with intermittent cooling (91 %). This clear preference was surprising, as earlier studies (Van Veen, 2016 and Chapter 5.1) have shown very different results with almost half the study participants disliking the cooling. The difference in the outcomes of these investigations is surmised to be due to the fanning intensity because in this study, there was no increase in the airflow towards the body.

Summarizing, RQ 2.2 can be answered as follows:

> A 2.2: Both the 2-minute and the 4-minute stimulus of 15 °C cold air reduce subjective fatigue as shown by the significantly lower KSS ratings. The 4-minute stimulus also increases physiological activation in the short term as seen from significantly increased pupil diameters and decreased HR. The 2-minute stimulus did not affect the pupil diameter and HR was only affected at the end of the stimulus.

One limitation of this particular study was that both COOL2 and COOL4 condition were tested within one drive instead of separate drives. Separate drives would have more equivalent starting conditions before cooling was applied, however, at the expense of a longer total duration of the study, because another 15-minute pre-cooling and 6-minute post-cooling time would have been required for a third drive. The fact that there was no significant difference between the KSS ratings after 33 minutes of the CONT drive and the COOL drive indicated that fatigue redevelops within 6 minutes while driving. This observation also means that the effect of the thermal stimulus does not have a longer lasting impact on fatigue. While the finding that arousal lasts for less than 6 minutes after a cooling stimulus, contributes to the understanding of fatigue, investigating how to potentially prolong these effects would be interesting for future research.

5.2.4 Conclusion

In the above study two different thermal stimuli (15 °C) of different durations, 2 minutes and 4 minutes, were tested in terms of their effect on subjective fatigue and on physiological indicators of fatigue. A driving simulator study was conducted in a within-subject design, including a drive in which the stimuli were applied and a control drive without cooling. An analysis of the objective and subjective responses of the study participants indicated that both stimuli caused psychophysiological arousal and decreased fatigue. According to participants' KSS ratings the drivers felt more awake after the cooling. Also the study's results on physiological measures showed that short-term cooling at 15°C was a countermeasure against passive fatigue while driving in monotonous road conditions. It could also be concluded from the collected data that thermal stimulation only had a short-lasting effect on fatigue.

A superior performance of the 4-minute stimulus can be derived from a longer effect on the physiological data as well as even lower subjective fatigue ratings.

With regard to successful fatigue management, the effectiveness of thermal stimulation is limited because the stimulus only has a short-term effect on the drivers' wakefulness. This observation is crucial because the stimulus would have to be applied repeatedly to cause a prolonged activation of driver and it is uncertain that the stimulus would still have an effect after a certain amount of repetitions.

5.3 Effect of 4-Minute Leg Cooling[5]

In the studies of Chapter 5.1 and 5.2, thermal stimulation of the upper body, including the face, has been shown to mitigate driver fatigue in the short term. Similarly, the hand cooling employed in the simulator study of Van Veen (2016) showed that this treatment invoked physiological arousal, which indicated sympathetic activation. Although, facial and hand cooling showed awakening effects, it had a negative impact on drivers' comfort ratings. Especially in the first cooling study of this thesis, in which a high

---

[5]   This chapter is based on a previous publication: Schmidt, E., Dettmann, A., Decke, R., & Bullinger, A. C. (2017c). Cold legs do not matter – investigating the effect of leg cooling to overcome passive fatigue. *Postersession Proceedings of the Human Factors and Ergonomics Society Europe Chapter 2017 Annual Conference.* Rome, Italy: The Human Factors and Ergonomics Society, Europe Chapter.

fanning intensity was applied, the participants commented, that the stimulus felt too cold (Chapter 5.1). Therefore, it is interesting to investigate the effect of thermal stimulation of body parts other than the upper body.

Inspired by water treading and its physiological effects (Ganz, 2016), an interest in leg cooling as a fatigue countermeasure led to the next RQ:

> RQ 2.3: Which effect does 4-minute leg cooling with 15 °C air has on subjective fatigue as well as on physiological measures?

This chapter is organized as follows: Chapter 5.3.1 contains the method applied in the study. Next, the results of subjective and objective measures of the different groups are reported in Chapter 5.3.2 and discussed in Chapter 5.3.3 and the conclusions about the effectiveness of leg cooling against passive fatigue are drawn in Chapter 5.3.4.

5.3.1 Method

Chapter 5.3.1.1 refers to of the technical setup in which the study took place. The study design is explained in Chapter 5.3.1.2, followed by the sample description in Chapter 5.3.1.3. The dependent variables collected in the study are listed in Chapter 5.3.1.4. Last, the data analysis is described in Chapter 5.3.1.5.

5.3.1.1 Setup

For this study in November 2016, the BMW i3 apparatus was used in the flat-screen simulator shown in Chapter 3.2 and 3.3. The cold air was generated with the car's AC and could be controlled remotely at the investigator's desk (see Figure 7).

5.3.1.2 Study Design

The study employed a between-subject design with two groups. Both groups completed a 5-minute familiarization drive and a 30-minute monotonous highway drive with no traffic, similar to the tiring drive in Chapter 4.1.

In the control group (CONT), the cockpit temperature was maintained at 24 °C for the entire drive. In the treatment group (COOL), the climate changed from the control settings between minute 20 and 24 (see Figure 26).

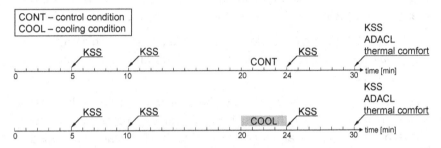

Figure 26: Test drives with marked times of verbal assessment of fatigue and cooling period
Source: Own illustration

In the 4 minutes where cooling was performed, the temperature dropped from 24 °C to 15 °C. The cooling duration of 4 minutes was chosen based on the results of the study in Chapter 5.2, which compared different durations in terms of their effectiveness against fatigue. At all times, the fan intensity was set to 30 % and the airflow was directed to the legroom of the car. The reason for directing the stimulus toward the drivers' legs and the associated physiological responses are described in Chapter 3.4.

5.3.1.3 Participants

Each group in the between-subject design consisted of 21 participants. Details about the requirements for participation are described in Chapter 3.5. A description of the demographic characteristics of the groups is presented in Table 13. A Mann-Whitney-U test showed that there was no difference in mean age between the two groups (Table 13). There was also no significant difference in sleep durations the night before the experiment between the groups (Table 13). Differences in sleep duration and fatigue rating after the familiarization drive between the groups would have skewed parts of the subsequent analysis. In this case, the KSS ratings recorded at the end of the drive would not only have been possibly affected by the cooling, but also by a higher immanent fatigue in one group or circadian effects.

Table 13:   Demographic description of sample size
Source: Own table

Parameter	Group 1	Group 2	Mann-Whitney-U test
group size	n_1=21	n_2=21	
age [y] (M, SD)	30.2 ± 12.0	26.1 ± 6.4	U=489, p=.35
age [y] (minimum, maximum)	20, 66	21, 49	
number of females per group	2	5	
sleep duration [h] in the night before the experiment (M, SD)	6.78 ± 0.83	6.95 ± 0.79	U=389.5, p=.57
KSS after familiarization drive (M, SD)	4.14 ± 1.46	3.90 ± 1.70	U=477, p=.52
condition	control (CONT)	leg cooling (COOL)	

5.3.1.4 Dependent Variables

During the experiment, the participants completed several questionnaires. They answered the KSS (Åkerstedt and Gillberg, 1990, translated into German by Niederl (2007)) before and after the familiarization drive as well as after the 30-minute test drive. Additionally, they responded to the KSS during the test drive after 5, 10 and 24 minutes of driving via a microphone-speaker system without pausing the drive (see Figure 26). Furthermore, the drivers filled out the ADACL (Thayer, 1989, translated into German by Imhof (1998)) after both familiarization and test drive. Last, the groups rated their thermal comfort with the Bedford scale (Bedford, 1936), where it was specified for the COOL group to rate their comfort during the coolest part of the drive. Last, the drivers of the COOL group were asked whether they liked the cooling and whether they perceived it would reduce fatigue like in Chapter 5.1 and 5.2 (see A2.3, p. 188).

Physiological data were also collected in the study. Details about the recording of the physiological data are described in Chapter 3.6.

5.3.1.5 Data Analysis

For details about the physiological data filtering and analysis, please see Chapter 3.7. The hypotheses for the between-subjects tests were the following: there would be no difference in KSS values between the groups for the self-assessments provided after 5 and 10 minutes of driving because in the first 20 minutes of the drive, the experimental conditions for both groups were identical. After 20 minutes of driving, it was hypothesized that fatigue in the COOL group would be lower, since earlier studies could show an activating effect of thermal stimulation (Janský et al., 2003). Therefore, one-sided significance tests were used for the means of KSS values given after 24 and 30 minutes of driving.

The significance tests for the comparison between groups according to the ADACL were one-sided because it was hypothesized that both positive and negative deactivation, namely "calmness" and "tiredness", would decrease after leg cooling and that positive and negative activation, "energy" and "tension", would decrease.

Mann-Whitney-U tests were used for the comparison of non-normally distributed data from different groups. For analyses of the fatigue induction within single groups, Wilcoxon-tests were used.

5.3.2 Results

In the following, the results on the subjective assessments of fatigue, activation and thermal comfort are presented in Chapter 5.3.2.1. Chapter 5.3.2.2 contains the results of the physiological measures.

5.3.2.1 Subjective Data

First, the results of the subjective questionnaires are presented. Table 14 lists the means and standard deviations of the KSS values for each group at different times of the highway drive. The fatigue increase within the CONT group was significant from 5 to 10 minutes (Z=-2.3, p=.03), 10 to 24 minutes (Z=-2.6, p=.01) and from 24 to 30 minutes (Z=-2.2, p=.03). In the

COOL group, the increase in fatigue was only significant from 5 to 10 minutes (Z=-3.6, p<0.001) and from 24 to 30 minutes (Z=-3.1, p=.002). The increase in subjective fatigue from minute 10 to 24 was non-significant (Z=-1.9, p=.06). The increase in self-reported fatigue in the CONT group and for the most part in the COOL group, showed, that the driving task induced passive fatigue.

Table 14 lists the significance test results which support that the fatigue ratings of the CONT and COOL group are indifferent after 5 and 10 minutes of driving. As hypothesized, the mean fatigue rating in the COOL group was significantly lower compared to the CONT group directly after the thermal stimulus (at 24 minutes). However, 6 minutes after the stimulus, the KSS values in the COOL group were not significantly lower than in the CONT group.

Table 14:   KSS rating (M and SD) at different times of the drive and significance test results
Source: Own table

Time of self-assessment	5 min	10 min	24 min	30 min
KSS (M, SD) CONT (n=21)	5.43 ± 1.72	5.91 ± 1.51	6.76 ± 1.64	7.24 ± 1.67
KSS (M, SD) COOL (n=21)	4.62 ± 1.63	5.29 ± 1.87	6.05 ± 1.40	7.05 ± 1.69
Mann-Whitney-U test	U=511, p=.14 (two-sided)	U = 497, p =.25 (two-sided)	U=520, p=.04 (one-sided)	U=466, p=.36 (one-sided)

Using the results of the ADACL, the means for positive and negative activation and deactivation were compared (Table 15). In general, the mean activation ratings were low after the monotonous drives (between 1.44 and 1.63 on a scale from 1 to 4), whereas the deactivation ratings were high. As Table 15 shows, none of the hypotheses could be confirmed based on the non-significant test results of the Mann-Whitney-U test.

Table 15:   ADACL results (M and SD) and significance test results between means
Source: *Own table*

Psychological construct of ADACL		M, SD (based on 1-4 scales)	Mann-Whitney-U test (one-sided)
positive activation	energy CONT	1.57 ± 0.40	U=512, p=.94
	energy COOL	1.44 ± 0.58	
positive deactivation	calmness CONT	3.10 ± 0.64	U=395, p=.93
	calmness COOL	3.35 ± 0.54	
negative deactivation	tiredness CONT	3.16 ± 0.46	U=459, p=.43
	tiredness COOL	3.09 ± 0.61	
negative activation	tension CONT	1.63 ± 0.66	U=499, p=.89
	tension COOL	1.38 ± 0.43	

Figure 27 shows the thermal comfort of both groups measured with the Bedford scale. The CONT group rated their thermal comfort on average with 1.3, whereas 1 means "comfortably warm" and 2 means "too warm". The mean thermal comfort of the COOL group was -1.0 which means "comfortably cool". The ratings of the COOL group were significantly lower than the ones given by the CONT group (U= 641, p<.001).

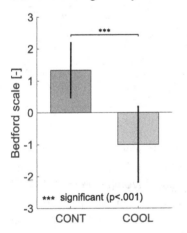

Figure 27:   Thermal comfort ratings (M and SD) of the CONT and COOL group
Source: *Own illustration*

The answers of the COOL group about their liking and perceived effectiveness of the cooling are listed in Table 16. Of the 21 drivers, only 13 indicated that they liked the cooling of the legs, however, 18 (86%) thought that the stimulus reduced their fatigue.

Table 16: Responses of the participants in the COOL group
Source: Own table

Responses of the COOL group about impressions of the cooling of the legs [number of people]	Yes	No
Did you like the cooling?	13	8
Do you think that the cooling helps against fatigue?	18	3

5.3.2.2 Physiological Parameters

ECG data, from which HR and SDNN were retrieved, was of usable quality for $n_{ECG, CONT}$=19 participants of the CONT group and for all of COOL group ($n_{ECG, COOL}$=21). The group averages of HR over time are visualized in Figure 28.

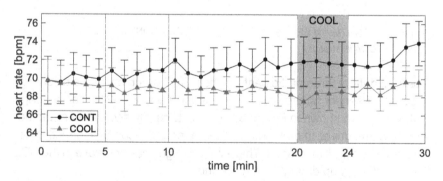

Figure 28: Heart rate (M and SE) over the course of the drives CONT and COOL
Source: Own illustration

There was no significant difference in the HR between the groups at any point in time. Non-overlapping standard error bars do not necessarily mean

that the difference is significant. Instead significance is defined by the result of the Mann-Whitney-U test. The times of verbal fatigue assessment
(5, 10 and 24 minutes) as well as the cooling period are highlighted in the
diagram with vertical lines.

The average values of the SDNN for both groups are shown in Figure 29.
In both groups, the SDNN was increasing over time. A comparison of the
first and last value of the mean SDNN showed that the HRV significantly
increased within each group (one-sided Wilcoxon test: Z=-1.67, p=.047 in
CONT and Z=-2.00, p=.02 for COOL). There were no significant differences between the CONT and the COOL group at any point in time.

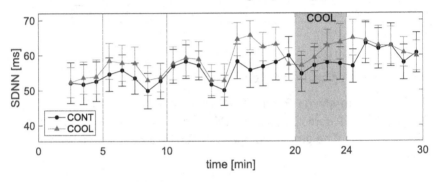

Figure 29:   SDNN (M and SE) over the course of the drives CONT and COOL
Source: Own illustration

The SCL data of three participants of the CONT group had to be discarded
because of sensor failure ($n_{SCL, CONT}$=18 and $n_{SCL, COOL}$=21). The average
SCL values are shown in Figure 30. The SCL rapidly decreased at the beginning of the drives in both groups. Moreover, the mean SCL directly after
verbal assessment increased. There was no significant increase in the SCL
in the COOL group during the leg cooling.

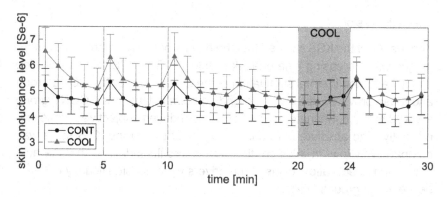

Figure 30: Skin conductance (M and SE) over the course of the drives CONT and COOL
Source: *Adopted from Schmidt et al., 2017c*

The measurements of the pupil diameter of three participants of the CONT group and one participant of the COOL group had to be removed from the dataset because of inadequate tracking accuracy, resulting in $n_{pupil, CONT}=18$ and $n_{pupil, COOL}=20$. Similar to the SCL measurements, the pupil diameter decreased over time and increased directly after the verbal fatigue assessments in minute 5, 10 and 24 (Figure 31). There were no significant differences between the two groups in pupil diameters, as Mann-Whitney-U tests showed.

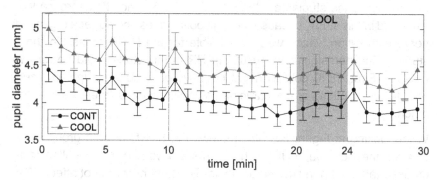

Figure 31: Pupil diameter (M and SE) over the course of the drives CONT and COOL
Source: *Adopted from Schmidt et al., 2017c*

5.3.3 Discussion

The results of the KSS showed that the drivers perceived significantly more fatigue with increasing time-on-task (Table 14). The KSS ratings provided directly after the thermal stimulation of the drivers' legs in the COOL group and those provided at the corresponding point in time in the CONT group showed that the 4-minute thermal stimulus reduced subjective fatigue significantly. The absolute difference, though, of 0.7 on the KSS scale was comparatively low. In addition, the ADACL results indicated that the activation and deactivation levels of the drivers were not significantly different between the groups (Table 15).

Furthermore, only 62 % of the drivers stated that they liked the cooling (Table 16). The people disliking it, stated that they did not like that the cooling was concentrated on the legs and that it was too cold. This aligns with the Bedford scale ratings that were negative in the COOL group and significantly different between the CONT and the COOL group (Figure 27).

The HR indicated that the cooling did not have an effect on the physiological fatigue of the drivers (Figure 28). In the case of leg cooling, it was expected, that HR would increase because this was seen in the laboratory study of Janský et al. (2003). In this study, a significant increase in HR when the lower legs were immersed in a 12 °C water bath was recorded. In contrast to the studies of Chapters 5.1 and 5.2, no HR decrease was expected in this study, because the leg cooling does not trigger a trigeminal nerve stimulation, which was responsible for the HR decrease in those studies. The non-significant results in HR in this study could be for two reasons. First, 15 °C cold air might not have been cold enough and, second, the leg clothing of the participants could have insulated the legs too much.

When looking at Figure 29, which shows the SDNN of both groups, one can see the increase in HRV. This shows a physiological indication of increased fatigue among the participants. The treatment did not affect HRV, as was also the case with upper body cooling in the studies of Chapters 5.1 and 5.2.

Both SCL and pupil diameter decreased over the course of the drive (Figure 30 and Figure 31). As in earlier studies (Chapters 5.1 and 5.2), the verbal assessment of the drivers' state with the KSS caused an increase in both SCL and pupil diameter. In contrast to the earlier studies, these parameters did not increase during leg cooling. This was unexpected, as this study applied the same stimulus as in the study of Chapter 5.2, namely 15 °C for 4 minutes. The only difference was, that the cold air was directed to a different body part. The lack of physiological activation in this study was probably due to the comparatively low amount of thermal cold receptors in the skin of the legs (see Table 4 in Chapter 2.2.1).

The significant decrease in subjective fatigue ratings was contradictory to the lack of effects on the physiological parameters, which requires some further explanation. One possible reason for this contradiction is that the drivers were subject to a placebo effect, since they clearly felt the cold air around the legs, which could be proved by their thermal comfort ratings (Figure 27). Hence, the lower fatigue ratings were probably due to the drivers' belief in the treatment, rather than to the cooling property itself.

Since the physiological parameters were not affected by the leg cooling, RQ 2.3 can be answered as follows:

> A 2.3: Leg cooling with 15 °C cold air for 4 minutes does not have an effect on the physiological measures HR, SCL and pupil diameter, indicating that no SNS activation occurred. The treatment had an effect on subjective fatigue, as the KSS ratings provided by the two experimental groups were significantly different, however, this is likely due to a placebo effect, considering that no physiological activation was measured.

In order to achieve fatigue-mitigating effects by means of leg cooling, the temperature of the air that directly surrounds the legs has to be lower. This can be substantiated with the study of Janský et al. (2003), who could show physiological effects, indicating an SNS activation, by means of leg cooling in a 12 °C water bath.

The study results may have been skewed by the fact that participants wore pants during the experiment and that the legs were consequently partly insulated (see the recruitment information in Chapter 3.5). Shorts may have resulted in successful fatigue mitigation, however, the acceptance of the treatment would have been worse due to low thermal comfort. Already in this study, in which the lower legs were covered, 38 % of the drivers stated that cooling as uncomfortable (Table 16).

5.3.4 Conclusion

The above study addressed the effect of thermal stimulation of the drivers' legs on subjective and physiological fatigue as well as on thermal comfort. The results of the simulator study showed that there was no valid indication of fatigue mitigation through the thermal stimulus on the drivers' legs as the physiological indicators were not affected.

Cooler temperatures and the exposure of uncovered lower legs might have caused an SNS activation that might be reflected in physiological parameters. Nevertheless, a reduction in temperature around the legs in the driving context, is not a good countermeasure overall, as the percentage of drivers disliking the cooling is relatively high already with 15 °C. Summarizing, leg cooling at 15 °C as a fatigue countermeasure is not recommendable as it is neither effective nor comfortable.

5.4 Effect of Thermal Stimuli – a Comparative Analysis

While Chapters 5.1, 5.2 and 5.3 dealt with the report of the single driving simulator studies and the gained insights of those, this chapter contains the comparison of the results of the three driving simulator studies. This comparison is done to provide a more general answer to RQ 2:

> RQ 2: Which effects do short-term thermal stimuli have on the drivers' passive fatigue?

This chapter is organized as follows: The method is presented in Chapter 5.4.1, followed by the report of the results in Chapter 5.4.2. The results are discussed in Chapter 5.4.3 and, last, conclusions about the effect and comfort of the different stimuli are drawn in Chapter 5.4.4.

5.4.1 Method

The three driving simulator studies of Chapters 5.1, 5.2 and 5.3 yielded different results in terms of effectiveness against fatigue and perceived thermal comfort. While it was possible to answer the three specific subordinate RQs 2.1, 2.2 and 2.3 by means of comparing COOL and CONT conditions, further methods for data analysis helped to respond to RQ 2 and draw more general conclusions on the effect of thermal stimulation by means of comparing the results.

The following Chapter 5.4.1.1, 5.4.1.2 and 5.4.1.3 refer to the details on study setup, design and participants, respectively. Chapter 5.4.1.4 deals with the relevant dependent variables for the data analysis which is described in Chapter 5.4.1.5.

5.4.1.1 Setup

For a detailed description of the setups please see Chapter 5.1.1.1 (p. 69), 5.2.1.1 (p. 87) and 5.3.1.1 (p. 101).

5.4.1.2 Study Design

Details of the study designs of the different studies are provided in Chapter 5.1.1.2 (p. 69), 5.2.1.2 (p. 87) and 5.3.1.2 (p. 101).

5.4.1.3 Participants

A description of the study participants is provided in Chapter 5.1.1.3 (p. 71), 5.2.1.3 (p. 90) and 5.3.1.3 (p. 102).

5.4.1.4 Dependent Variables

As explained in Chapter 1.1, there is a trade-off in the design of suitable thermal stimulation to both reduce fatigue and maintain the driver's comfort. Therefore, dependent variables are needed that capture the effect of the thermal stimulation on both fatigue indicators and comfort.

Effect on fatigue indicators

The dependent variables are the difference in KSS ratings, the presence of an effect in subjective KSS ratings by means of the tested stimulus, and the presence of effects in HR, SCL and pupil diameter caused by the thermal stimulation. The presence of the effect is binary coded, i.e., a significant effect is present or not. The effects on HRV measures, such as SDNN and RMSSD are not included in the analysis because studies showed, that the effect of cooling is only short lived and therefore HRV measures that require a raw ECG signal window of several minutes do not reflect brief changes in HR. In addition, the responses to the question "Do you think that the cooling helps against fatigue?" are compared for the different stimuli as another subjective measure of perceived effectiveness.

While the KSS, HR, SCL and pupil diameter effects can be used as indicators for the fatigue-mitigating effect of thermal stimulation, different measures are needed to assess the effect of cooling on the passengers comfort.

Effect on comfort

For this, the number of positive feedback responses to the question "Did you like the cooling?" were used as an indicator for comfort. The results of the Bedford scale (Bedford, 1936) are not included in the analysis because the thermal comfort rating reflects only a snapshot of comfort at the time of maximum cooling. Instead the statement to like the cooling captures an overall impression on the cooling.

5.4.1.5 Data Analysis

The data that reflected the effectiveness of thermal stimulation as a measure against passive fatigue, i.e., the absolute difference in the KSS ratings, the binary coded presence of effects in KSS, HR, SCL and pupil diameter for the different treatments, were compared qualitatively against each other. The perceived-effect rates, i.e., the number of people who thought that cooling helps against fatigue, in the three studies applying different stimuli will be compared using a Chi-square test.

Similarly, the positive-feedback rates, i.e., the number of people liking the cooling, in the three studies applying different stimuli will also be compared by means of a Chi-square test.

5.4.2 Results

A summary of the results of the dependent variables is shown in Table 17. As a reference, Table 17 also includes a brief a recap of the stimulus description.

The 6-minute stimulus of an increased cold airflow of 17 °C directed to the upper body yielded significant effects in the subjective KSS ratings and all relevant objective parameters. The 15 °C cold air directed to the upper body did not yield an increase in SCL for both the 2-minute and the 4-minute duration. The 2-minute stimulus also failed to trigger an increase in pupil diameter. The study on leg cooling at 15 °C for 4 minutes revealed only subjective effects of increased wakefulness, however, all objective indicators of fatigue were unaffected by the thermal stimulus. The Chi-square test of the perceived-effect rates yielded no significant effect of the treatment on the perceived effect of the cooling ($\chi^2(2)$= 4.68, p=.096).

The Chi-square test of the positive-feedback rates revealed a significant effect of the treatment on the liking of the cooling ($\chi^2(2)$= 6.05, p=.048).

Table 17:   Summary of the results of the three studies on thermal stimulation
Source: *Own table*

Stimulus characteristics		Chapter 5.1	Chapter 5.2		Chapter 5.3
stimulus temperature		17 °C	15 °C		15 °C
airflow compared to CONT condition		increased	constant		constant
body part		upper body	upper body		legs
duration		6 minutes	2 minutes	4 minutes	4 minutes
Results					
fatigue indicators	mean difference in KSS	1.3	0.5	0. 6	0.7
	effect in KSS	x	x	x	x
	effect in HR	x	x	x	
	effect in SCL	x			
	effect in pupil diameter	x		x	
	number of drivers perceiving effect	30 (88 %)	33 (100 %)		18 (86 %)
	number of drivers not perceiving effect	4	0		3
comfort	number of drivers liking the cooling	20 (59 %)	28 (85 %)		13 (62 %)
	number of drivers disliking the cooling	14	5		8

5.4.3 Discussion

As stated in Chapter 1.1, this thesis focuses on the discomfort-related aspects of passive driver fatigue. This means that thermal stimulation on the one hand, needs to mitigate fatigue and hence the associated discomfort and on the other hand, thermal stimulation needs to avoid discomfort arising from thermal sensation of the driver. The results will therefore be discussed in terms of these two aspects.

Effect on fatigue indicators

First, the subjective effectiveness of the different treatments was compared. The stimulus used in the first study, which consisted of 6-minute upper body cooling at 17 °C with increased airflow, yielded the absolute difference of 1.3 in the KSS ratings between the cooling and control condition which is higher compared to the other studies, in which the difference ranged from 0.5 to 0.7. In all studies, the difference was statistically significant. The first study probably yielded a larger difference because of the increased fan intensity that came along with a colder perception of the 17 °C air through wind chill effects, and also with a secondary auditory and tactual effect. The number of drivers perceiving that cooling helped against fatigue are similarly high in all studies and ranges from 86 % for the 4-minute leg cooling to 100 % for the 2- and 4-minute upper body cooling. This observation suggests that participants rated the thermal stimulation as helpful using low thresholds. Participants probably thought of the stimulation as helpful as soon as they perceived only little relief from the strain of monotonous driving. Therefore, also the second (Chapter 5.2) and third study (Chapter 5.3) yielded high perceived-effect rates, even though the KSS-rating decrease was not as much as for the first study (Chapter 5.1).

Second, the objective effectiveness of the different treatments was compared. In the study applying 17 °C air at increased volume flow rates, HR, SCL and pupil diameter were significantly affected, in a way that indicated an SNS activation. As explained above, the increased fan intensity caused a secondary auditory and tactual effect. It can be argued that this secondary effect only affected SCL and pupil diameter in the first minute (see discussion in Chapter 5.1.3). Still, this treatment yielded more effects in physiological data than the 4-minute upper body cooling at 15 °C with constant low airflow, during which SCL was not affected. Even less effects were yielded by the 2-minute stimulus because it failed to affect the pupil diameter. The reason for the presence of more physiological effects of the 17 °C stimulus compared to the 15 °C stimulus is very likely due to the additional wind chill in the 17 °C condition which was caused by the air movement. Air movement (in addition to air temperature, radiant temperature and humidity) is an important factor in defining the thermal environment (Chapter 2.2.1).

The 4-minute long leg cooling at 15 °C with constant airflow did not yield any objective indication of reduced fatigue as the physiological measures were not affected. The difference in the effect of the 4-minute stimulus applied to the legs (Chapter 5.3), compared to the upper body (Chapter 5.2), can be reasoned with the distribution of cold thermal receptors (see Table 4 in Chapter 2.2.1), which are more densely distributed around the neck and face. Stuke (2016), for example, makes use of this characteristic of thermal perception in his research about vertical cooling systems for cars.

Summarizing, both the 6-minute cooling at 17 °C and the 2- and 4-minute cooling at 15 °C of the upper body proved subjectively and objectively to be mitigating fatigue, whereas 4-minute leg cooling at 15 °C failed to show an objective indication of fatigue mitigation. As discussed in Chapter 5.3.3, leg cooling may cause physiological arousal, if the stimulus temperature was lower and if the legs were not covered by clothing as seen in Janský et al. (2003) who found physiological effects through leg cooling in a 12 °C water bath.

Effect on comfort

Finally, the drivers' impressions of the cooling were compared. In terms of likeability, the 2-minute and the 4-minute stimuli at 15 °C for the upper body were superior as the significant Chi-square test showed. In this study, 85 % of the drivers liked the cooling. This is higher compared to the 6-minute stimulus at 17 °C, which only 59 % of the drivers liked (Table 11, p. 74). Even though the temperature was higher in this study, the impressions of the cooling are worse most likely, because the drivers did not like the high fan intensity and the long duration. In case of the leg cooling, only 62 % of the drivers liked the treatment. The lower positive-feedback rate of the 15 °C leg cooling compared to the 15 °C upper body cooling, may be due to the phenomenon that people in general prefer to have warm legs and a cool head while driving rather than the other way round (Vetter et al., 2003).

In conclusion, the three studies allow for answering the RQ 2:

> A 2: Short-term thermal stimuli applied to the upper body reduce passive driver fatigue in the short term as both subjective data indicate a mitigation of fatigue and physiological patterns of SNS activation can be measured. Using the same temperature, thermal stimulation of the drivers' legs does not prove to reduce passive fatigue. Drivers' comfort is least disturbed when constant low airflow is directed towards the upper body.

Taking all results into account, the 4-minute thermal stimulus of 15 °C directed to the upper body is recommended as a fatigue countermeasure because it mitigated subjective and physiological fatigue, but has minimal negative effect on comfort. Therefore, the last study (see Chapter 7) applies this countermeasure.

5.4.4 Conclusion

When comparing the study results of Chapters 5.1, 5.2 and 5.3 with each other, several conclusions about thermal stimulation as a fatigue countermeasure can be drawn.

First, both studies on upper body cooling (Chapter 5.1 and 5.2) revealed physiological indications of reduced fatigue as well as significantly lower KSS ratings, while the study on leg cooling (Chapter 5.3) failed to substantiate the lower KSS ratings with corresponding physiological measurements. Therefore only the first two studies fulfilled the requirement of effectiveness against fatigue and therefore the mitigation of discomfort arising from fatigue.

Second, the comparison of the positive-feedback rates showed superiority of the 2-minitue and 4-mintue upper body cooling applied in Chapter 5.2 and inferiority of the 4-minute leg cooling applied in Chapter 5.3. Leg cooling was perceived as uncomfortable and it goes against the general preference of warm legs and a cool head while driving (Vetter et al., 2003). The 2- and 4-minute upper body cooling at 15 °C were also superior to the 6-

minute cooling at 17 °C in terms of comfort because the airflow was kept constant.

Because of the superiority of the 4-minute upper body cooling at 15 °C in terms of comfort and its effect on KSS, HR and pupil diameter which are symptoms of an SNS activation, this setting is recommended for further studies.

Whereas the above studies provided insights into whether short-term cooling was effective against passive fatigue, they did not provide the answer to whether repeated cooling would still be effective. There is reason to assume that the change of thermal conditions is especially stimulating for the driver and an effect on fatigue may be visible with every change of thermal conditions. Based on the studies of the orienting response with repeated stimuli (Goldwater, 1972; Graham and Clifton, 1996), it can also be assumed that immediate physiological responses, like pupillary dilatation and HR decrease, will be less apparent with repeated cooling. Therefore, it is important to study the effect of repeated cooling on the driver's subjective and physiological fatigue (see Chapter 7) to develop most effective countermeasures. Knowledge about the effect of repeated treatments is especially important, as the results above have shown that the wakening effect is only short-lived. If a continued effect of repeatedly applied thermal stimuli on passive fatigue could be proven, this may create new opportunities to increase alertness for periods longer than just a few minutes.

6 Detection of Fatigue Based on Physiological Measurements[6]

The following chapter deals with the analysis of data collected during several driving simulator studies. The analysis was used to generate a fatigue model which takes objectively measureable data as input signals and outputs an objective fatigue assessment.

Objective fatigue assessment is essential for studies aiming at driver fatigue mitigation, in order to:

- trigger the thermal stimulus at an appropriate time
- evaluate the effect of the stimulus

Subjective fatigue ratings cannot meet these requirements because these can only be provided at discrete times, preventing a continuous analysis of fatigue. Furthermore, listening to and answering questions has an awakening effect on the driver, which is often undesirable in driver fatigue research.

As described in Chapter 2.1.4, there exists a variety of algorithms, which classify different levels of sleepiness. Most of these algorithms have been developed, however, using data from sleep-deprived drivers, and are only valid for detecting SR fatigue. Therefore, a regression model that can be used as a tool for continuously evaluating passive driver fatigue caused by monotonous driving in simulator studies is described in this chapter.

---

[6] This chapter is based on a previous publication: Schmidt, E., Ochs, J., Decke, R., & Bullinger, A. C. (2017b). Evaluating drivers' states in sleepiness countermeasures experiments using physiological and eye data - hybrid logistic and linear regression model. *Proceedings of the 9th International Driving Symposium on Human Factors in Driver Assessment, Training and Vehicle Design* (pp. 284-290). Manchester Village, VT: University of Iowa. https://doi.org/10.17077/drivingassessment.1648.

© Springer Fachmedien Wiesbaden GmbH, part of Springer Nature 2020
E. Schmidt, *Effects of Thermal Stimulation during Passive Driver Fatigue*,
Gestaltung hybrider Mensch-Maschine-Systeme/Designing Hybrid Societies,
https://doi.org/10.1007/978-3-658-28158-8_6

As the results of the correlation analyses of subjective fatigue ratings with objective parameters in Chapters 4.2.3 and 5.1.2.3 have shown, even the most correlated parameters have a correlation coefficient of under $|r|=0.35$. These moderate correlations do not allow for precise fatigue modeling if using a single input parameter. Therefore, it is worthwhile to investigate, whether combinations of different signals reveal patterns that are better indicators of fatigue. The key RQ is:

> RQ 3: At which accuracy can passive fatigue be detected based on physiological measures?

To answer this question, training data collection from three simulator studies and the selection of features indicating fatigue is described in Chapter 6.1. Second, the fatigue model, consisting of a cascaded logistic and linear regression model, is detailed with its prediction accuracy in Chapter 6.2. Next, the results are discussed in Chapter 6.3. Finally, the conclusions of the study are drawn in Chapter 6.4.

6.1 Method

The following chapters describe the studies' setup (Chapter 6.1.1), the design of the studies (Chapter 6.1.2), the included sample (Chapter 6.1.3), the dependent variables (Chapter 6.1.4) and how data were analyzed (Chapter 6.1.5).

6.1.1 Setup

All of these studies were performed with the BMW i3 apparatus (Chapter 3.3) in the curved-screen simulator (see Figure 5 in Chapter 3.2).

6.1.2 Study Design

The primary aim of the pilot experiments providing the data for this secondary data analysis, was the investigation of in-car countermeasures for critical driver states by means of sensory stimulation. The studies were part of a research project that dealt with the mitigation of passive driver fatigue or stress using different kinds of stimulation, such as climate, light, sound and scent. The studies were conducted by the author and comprised the same methodological design in terms of experimental setup (Chapter 3.2 and 3.3), recruitment of participants (Chapter 3.5), sensors (Chapter 3.6) and data analysis (Chapter 3.7) as the other studies of this thesis. Because the studies' primary aims are for the most part of no further importance of the thesis, a holistic description of all tested conditions is spared here. The only reason for utilizing some of the data collected in these additional simulator drives is that this increases the amount of labeled data for the generation of the fatigue model. Labeled data in this context mean physiological measurements that were recorded during a monotonous driving scenario for which subjective fatigue ratings exist.

While sparing a full description of all experimental conditions and the study designs of the additional simulator studies, at least the monotonous drives are described from which data were included in the secondary analysis in the following.

The first experiment included a 24-minute long monotonous drive, listed in Table 18. Participants answered the SSS via a microphone-speaker system without pausing the drive after 14 minutes and at the end of the drive. A countermeasure was applied between minute 20 and 23 that consisted of a change in the car's interior. This change consisted of a combination (COMB) of orange light from the car ceiling, scent, rhythmic sound and an increased fan intensity of the AC. The examiner started these in-car settings via remote control from the control room shown in Figure 6.

The second study that provided data for the modeling of fatigue comes from the cooling study detailed in Chapter 5.1. In this study, each subject drove two identical highway routes for 26 minutes. In one of the drives – the order was counterbalanced – cooling at 17 °C (COOL) was applied between minute 20 and 26. Participants evaluated their fatigue after 6, 16 and 26 minutes with the KSS.

Table 18:   Overview of simulator drives used for model development
Source: Schmidt et al., 2017b

Date	Sample	Fatigue intervention
study I December 2015	n=36 (25 ♂, 11 ♀), age 31.3 ± 9.8 y, (minimum 18, maximum 57)	COMB: combination of light, sound, scent and climate in minutes 20-23

SSS ... COMB ... SSS
0 ———————— 14 ———— 20 — 23 24 → time [min]

| study II February 2016 | n=44 (31 ♂, 13 ♀), age 33.0 ± 11.4 y, (minimum 21, maximum 59) | COOL: climate change in minutes 20-26 |

KSS ... KSS ... CONT ... KSS
0 —— 6 ———————— 16 ———————— 26 → time [min]

KSS ... KSS ... COOL ... KSS
0 —— 6 ———————— 16 —— 20 ———— 26 → time [min]

| study III August 2016 | n=42 (33 ♂, 9 ♀), age 30.7 ± 8.7 y, (minimum 22, maximum 52) | None, CONT: control condition |

CONT ... SSS
0 ———————————————————— 18 → time [min]

The third experiment included an 18-minute long drive with no intervention (CONT). At the end of the drive, the participants rated their subjective fatigue with the SSS.

Table 18 summarizes the dates and sample size of the studies, the durations of the highway drives, times of verbal fatigue assessment, used questionnaires and a description of the applied sensory stimuli.

All drives were highway drives with very little traffic in the style of the tiring drive described in Chapter 4.1.2. Based on the observations of the study in Chapter 4 which addressed the possibility to induce passive fatigue by means of traffic scenarios, it is concluded that 17 minutes of monotonous driving in a simulator is sufficient to evoke average SSS ratings of 4.4 which is between "somewhat foggy, let down" and "foggy, losing interest in remaining awake, slowed down".

6.1.3 Participants

For each of the three experiments 50 subjects participated. As in the other studies, BMW Group employees were recruited via e-mail lists (see Chapter 3.5) and voluntarily participated. All 150 subjects maintained their regular sleep schedule. Of those, n=28 datasets were excluded from analysis due to sensor failure or simulator sickness and the usable datasets was n=122. Details about the requirements for participation are described in Chapter 3.5. Table 18 shows demographic data of the sample population.

6.1.4 Dependent Variables

As mentioned in Chapter 6.1.2, two important types of dependent variables were collected during the experiments:

- subjective fatigue ratings – used as ground truth (classes)
- physiological measurements – used as predictor variables

The SSS and KSS were used for measuring subjective fatigue. The points in time at which fatigue was rated by the participants varied across the studies (see Table 18).

Physiological data were recorded throughout the drives. Details about the recording are described in Chapter 3.6.

6.1.5 Data Analysis

The recorded data were processed and analyzed as described in Chapter 3.7. Additional data preparation steps were required for the regression analyses which are described in the following. The regression model was developed with the software weka 3.6.13 which is a data mining tool developed by the University of Waikato, New Zealand (Hall et al., 2009). The different processing steps are visualized in Figure 32.

Predictor Variables – Previous correlation analyses (Chapters 4.2.3 and 5.1.2.3) showed that HRV and pupil diameter were most correlated to fatigue and were therefore included as inputs. Furthermore, weka 3.6.13 allows for quick training and testing with various inputs. These tests showed, that the SCL increases the classification accuracy which is why SCL was also included as an input. The frequency domain HRV measures LF, HF and the total power of the spectral density of 3-minute HR sequences were also chosen as predictor variables.

To increase the classification accuracies, the predictor variables were further transformed. Instead of using absolute values, the relative changes of variables compared to the first driving minute, as shown in equations 1 and 2, yielded better results. For the HRV measures (RMSSD, HF and total power) the third minute of the drive was used as a reference value (see equation 3) because the HRV values required a signal window of 3 minutes. The subscript "rel" in equations 1, 2 and 3 indicates the relative change of the parameters over time compared to their respective value at the beginning of the measurement. Furthermore, the exponents of SDNN, HF_{rel}, total power$_{rel}$, SCL_{rel} and diameter$_{rel}$ were adjusted (see Figure 32). This transformation increases the kurtosis in their respective distributions up to 9 times and reduces the overall variability of the parameters.

$$diameter_{rel}(t) = \frac{diameter(t) - diameter(1)}{diameter(1)}, \ t...time \ [min] \quad (1)$$

$$SCL_{rel}(t) = \frac{SCL(t) - SCL(1)}{SCL(1)}, \ HRV_{rel}(t) = \frac{HRV(t) - HRV(3)}{HRV(3)} \quad \begin{array}{c}(2),\\(3)\end{array}$$

Classes – The responses of the participants to the KSS and SSS were used for labeling the data. The corresponding predictor variables were the values of the transformed physiological parameters in the minute before the subjective rating (e.g. the KSS rating after 16 minutes was matched with the variables from the 16th minute). The classes "alert" and "fatigued" were formed from the subjective fatigue ratings: Data observations with KSS-values of 1, 2, 3 and 4 as well as SSS-values of 1, 2 and 3 formed class 1 – "alert". Data observations with KSS-values of 8 and 9 as well as

SSS-values of 6 and 7 formed class 2 – "fatigued". This way, a total data set of 171 observations from 88 different drivers was generated. The dataset included 85 observations of alert drivers and 86 observations of fatigued drivers. An amount of 207 observations with mid-range KSS and SSS values was removed from the training and testing.

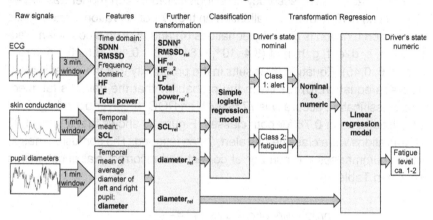

Figure 32: Signal processing for fatigue detection
Source: *Schmidt et al., 2017b*

Model – The detection of fatigue can be handled in two different ways: classification or regression. This is possible because the classes 1 and 2 do not only describe nominal classes (alert and fatigued) but can also serve as numeric values for the degree of fatigue, allowing for regression approaches. Regression approaches often fail to model the individual differences in prediction problems because the prediction is often approximating the mean of all labels. Therefore, a classification approach over regression was chosen in the first step to better distinguish the separate classes. To improve the sensitivity of the prediction results to external stimuli, a linear regression model with the inputs diameter$_{rel}$ and the class values was developed in the second step.

6.2 Results

After comparing several classification algorithms, random forest, neural network, decision trees, K-nearest neighbors, with the software weka 3.6.13, it was found that the logistic regression classifier performed best in terms of classification accuracy. The logistic regression model was developed using a 10-fold-cross-validation on the 171 observations. The logistic regression function is given by equations (4) and (5) with a coefficient vector (a, b, c, d, e, f, g, h, i) of (8.4· 10^{-5}, 0.81, -0.84, 0.16, -6.5·10^{-7}, 4.1·10^{-3}, 4.6, -6.6, 0.45). Equation (4) results in the probability that the driver is alert, whereas equation (6) results in the probability that the driver is fatigued. The classification accuracy is 77 %, with an ROC (receiver operating characteristic) area of 0.78 for both classes. Figure 33 shows that a total of 92 observations were classified as alert, 79 as fatigued. The confusion matrix, which summarizes the number of correct and incorrect classifications, is shown in Table 19.

$$probability\ of\ class\ "alert" = \frac{1}{1+exp(-x)} \tag{4}$$

$$x = a \cdot SDNN^2 + b \cdot RMSSD_{rel} + c \cdot HF_{rel} + d \cdot HF_{rel}^2 + e \cdot LF \tag{5}$$

$$+f \cdot total\ Power_{rel}^{-1} + g \cdot SCL_{rel}^3 + h \cdot diameter_{rel}^2 + i$$

$$probability\ of\ class\ "fatigued" = 1 - probability\ of\ class\ "alert" \tag{6}$$

The logistic regression model is not sufficiently accurate to model the slight fatigue mitigation of the driver through the countermeasures because the class is either 1 – "alert" or 2 – "fatigued". Hence, the model does not represent any intermediate state, such as slight reductions in fatigue due to in-car stimulation.

Table 19: Confusion matrix of simple logistic regression

(digits represent the absolute amount of observations)

Source: Schmidt et al., 2017b

Classified as →	1	2
1	69	16
2	23	63

Table 20: Confusion matrix of hybrid logistic and linear regression

(digits represent the absolute amount of observations)

Source: Schmidt et al., 2017b

Classified as →	1	2
1	70	15
2	24	62

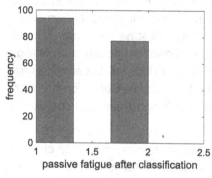

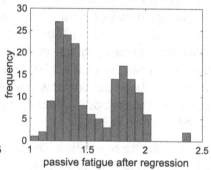

Figure 33: Histogram of predicted passive fatigue with simple logistic regression

Source: Schmidt et al., 2017b

Figure 34: Histogram of predicted passive fatigue with hybrid logistic and linear regression

Source: Schmidt et al., 2017b

Therefore, the model was improved by cascading a linear regression model after the logistic classifier, which can capture such intermediate states. The classification result is transformed to a numeric value and serves as an input for the linear regression model, along with the relative change in pupil diameter. The pupil diameter was chosen as an input because it is a very sensitive measure of sympathetic activation and hence able to replicate slight changes of driver activation. The linear regression model for fatigue was generated using a 10-fold-cross-validation. The regression function is given by equation (7) with a coefficient vector (j, k, l) of

(0.45, -1.5, 0.8). The correlation coefficient of the regression model is r=0.53, p<.001. The classification accuracy of the rounded numeric fatigue level is still 77%. Table 20 shows the confusion matrix and Figure 34 the distribution of predicted fatigue levels according to the linear regression model.

$$fatigue\ level=j \cdot class+k \cdot diameter_{rel}+l \qquad (7)$$

A comparison of the distributions in Figure 33 and Figure 34 shows the effect of cascading the linear regression: The linear regression approximates the mean of all observations, which means that the two bars of Figure 33 move closer towards the mean of 1.5 in Figure 34. This step does not alter the classification accuracy as the sum of the diagonal elements of the confusion matrices in Table 19 and Table 20 stays the same. The cascading of the linear regression, however, allows for the detection of slight changes in fatigue induced by external stimuli. If the linear regression would have been performed without the logistic classifier beforehand, the distribution of predictions would peak at 1.5 and therefore, confuse fatigued and alert drivers.

To evaluate the sensitivity of the proposed regression models to changes in passive fatigue due to stimulation, the predicted fatigue of the two drives CONT and COOL of the simulator study II were compared (see Table 18). Figure 35 and Figure 36 show the mean fatigue levels of the 44 drivers for the simple logistic and the hybrid logistic and linear regression, respectively. It can be seen that fatigue increased over the course of both drives. The simple logistic regression model (Figure 35) failed to capture the slight reduction in fatigue due to the cooling treatment because there were hardly significant differences between the CONT and COOL condition in minutes 20 to 26. For the hybrid regression approach (Figure 36), the t-test results for each driving minute yielded statistically significant differences between the two conditions in the 21st (p=.003), 23rd (p=.049), 24th (p=.034), 25th (p=.023) and 26th (p=.001) minute in which cooling was applied. The graph also shows that there are drops in fatigue after minute 6 and 16, when the drivers responded to the KSS, which had an awakening effect as expected.

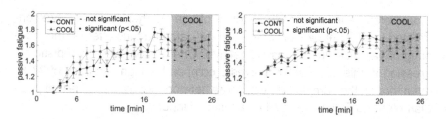

Figure 35: Predicted fatigue (M and SE) by
simple logistic regression
Source: *Schmidt et al., 2017b*

Figure 36: Predicted fatigue (M and SE) by
hybrid logistic and linear
regression
Source: *Schmidt et al., 2017b*

6.3 Discussion

The result of the analyses was a simple logistic regression model which was 77 % accurate at classifying passive fatigue using ECG, SCL and pupil diameter as input parameters. This accuracy is lower than the ones that have been reported in previous studies, as Table 3 (in Chapter 2.1.4) shows. Patel et al. (2011) for example described a neural network classifying sleepiness with 90 % accuracy based on drivers' ECG data. Friedrichs and Yang (2010) and Hu and Zheng (2009) reported accuracies of 83 % when differentiating three degrees of sleepiness based on camera or EOG data extracting eye features. Most of the algorithms listed in Table 3, however, have been developed using data from sleep-deprived drivers, hence these are detecting SR fatigue.

The work done by Igasaki et al. (2015) and Khushaba et al. (2011) is more comparable to this study because they reported on models that have been generated using the data of non-sleep deprived drivers. The logistic regression of Igasaki et al. (2015) based on HRV measures and respiratory features yielded 81 % detection accuracy. The sample size of eight male drivers aged between 21 and 23 years, though, is much smaller compared to the 88 drivers that provided the data for this analysis and therefore their results cannot be transferred to other samples. In this study, the larger and more heterogeneous sample, in terms of age and gender, may have contributed to a higher overall variance in the data and thus affected the modeling of variance due to passive fatigue. One reason, why the logistic regression in this study still achieves a comparably high accuracy of 77 %, even though the sample size was more heterogeneous, could be that pupil diameter and SCL were used as additional inputs.

The model of Khushaba et al. (2011) performs better compared to the model of this study. They reported classification accuracies of 97 % for differentiating five degrees of fatigue. Their model was generated with data from 31 simulator drivers aged between 20 and 69 years. The reason for this very precise modeling of fatigue is probably that besides ECG and EOG, EEG data were also used as input variables.

Summarizing the presented results, the RQ 3 can be answered:

> A 3: The passive fatigue of car simulator drivers can be detected with a simple logistic regression with an accuracy of 77 % using ECG, SCL and pupil diameter as inputs. The proposed fatigue model achieves a fair classification accuracy, taking into account that the data were collected from non-sleep-deprived drivers. Better performing models with similar signal input requirements found in the literature were trained with data from sleep-deprived drivers. It is unknown whether these models can also predict passive fatigue as accurately.

Moreover, the hybrid logistic and linear regression model detected changes in fatigue, as Figure 36 shows, which were induced by thermal stimulation in the sample in study II (see also Chapter 5.1). The predicted values also reflected the awakening effect when the driver is answering the KSS. Though, when applying the model for the evaluation of the effectiveness of fatigue countermeasures, it is recommended to have a large sample size to overcome the imprecision of the prediction.

When comparing the classification results of the logistic regression with the results of the hybrid logistic and linear regression, there is no improvement in classification accuracy by cascading the linear regression in the hybrid model. This additional step, however, reproduces slight changes in fatigue by allowing for continuous fatigue values. Furthermore, this linear regression is strongly influenced by the relative change in pupil diameter, which is a very sensitive measure of activation.

As changes in the brightness cause pupillary restrictions unrelated to an increase in fatigue, those changes would skew both classification and regression results. Therefore, the proposed regression model will only perform well if the light settings are kept constant during the experiment. For this reason, the regression model is not suited to evaluate light as an intervening stimulant for fatigued drivers. In further work, it might be possible to take brightness-related changes of the pupil diameter into account, e.g., in form of a correcting factor or term. Pfleging et al. (2016) for example have demonstrated how to model task difficulty based on pupil diameter measurements under various lighting conditions.

Since the fatigue model was trained with the extreme KSS- and SSS-values, the reported accuracy can only be guaranteed for alert or fatigued ratings. For intermediate values (5, 6, 7 for KSS and 4, 5 for SSS) the logistic regression probabilities are close to an equal likelihood for both classes, increasing the risk of misclassifications.

6.4 Conclusion

This study dealt with the secondary data analysis of subjective fatigue ratings and physiological measurements recorded during a series of simulated drives. The aim of the analysis was the modeling of the subjective fatigue ratings based on using specific physiological measurements as input variables.

Using a simple logistic regression, the states "alert" and "fatigued" could be classified with an accuracy of 77 % using the input signals ECG, SCL and pupil diameter. A cascaded linear regression model furthe enabled the modeling of slight changes in fatigue – e.g. induced by thermal stimulation – using the pupil diameter and the result of the simple logistic regression model as input parameters.

Taking both the accuracy and sensitivity of the regression model into account, the regression model is a suitable tool for continuously evaluating passive fatigue due to monotony in driving simulator studies. The fatigue model can be considered to be more robust compared to those found in literature because of the large variety in training data, as data of 88 different drivers were used to develop and cross-validate the regression models, which is significantly higher compared to other presented models (see Table 3). The model can also serve as an objective measure for the effectiveness of countermeasures, such as in-car stimulants. This is another unique characteristic about the proposed method, as none of the existing fatigue models had a focus on detecting slight changes in fatigue caused by sensory stimulation.

In future research, the regression model should also be tested for different causes of TR fatigue, e.g., during automated driving. In further studies (see Chapter 7), the performance of the fatigue prediction for repeated and thermal stimuli will be tested.

7 Driver Vitalization Through Fatigue-Based Climate Control

In the previous studies of this thesis (Chapters 5.1 and 5.2), it has been shown that thermal stimulation of the upper body has a wakening effect on a passively fatigued driver. These stimuli, however, were applied after a pre-defined amount of time, in which the average driver subjectively perceives the onset of fatigue, according to the fatigue induction study in Chapter 4. As there are individual differences in the development of fatigue, it could have been that some participants felt very tired when cooling started, and might have benefited from an earlier start of cooling. Others may not have felt tired yet when cooling started, and they might have preferred a later start of the cooling. For the future use of this countermeasure, it is desirable that the stimuli are triggered when fatigue is detected.

Similar ideas were pursued by Desmond and Matthews (1997) who suggested two criteria for a fatigue countermeasure system. First, the system must provide a reliable indication of fatigue and second, there must be a stimulus upon the detection of fatigue in order to restore the performance of the driver. Another description of a fatigue management system is provided by Balkin et al. (2011) who defined that an ideal system measures fatigue while driving and intervenes as effectively and as often as needed when vigilance deficits are predicted.

Therefore, it is worthwhile to test a system, which detects fatigue – using the simple logistic regression model of Chapter 6 – and upon detection, communicates this to the driver and then provides thermal stimulation as a fatigue countermeasure. This study also allowed for the investigation of repeatedly applied identical stimuli, which had not been tested in any of the previous studies of this thesis. Furthermore, the study included different interaction strategies to start and announce the cooling. This is important to the future development of fatigue management systems, as the changes within the vehicle environment due to a countermeasure, on the one hand,

© Springer Fachmedien Wiesbaden GmbH, part of Springer Nature 2020
E. Schmidt, *Effects of Thermal Stimulation during Passive Driver Fatigue*,
Gestaltung hybrider Mensch-Maschine-Systeme/Designing Hybrid Societies,
https://doi.org/10.1007/978-3-658-28158-8_7

may be confusing to the driver if not announced properly. On the other hand, verbose announcements of the countermeasures may also be annoying to the driver.

The importance of the appropriate communication of automated assistance functions has already been described by Coughlin et al. (2009) and Reimer et al. (2009) within their framework of an "Aware Car" which would (a) detect the driver's state, (b) display that information to the driver and (c) offer in-car systems to refresh the driver (Coughlin et al., 2009). This suggests that the display of the information is essential. Knijnenburg et al. (2012) described in their work that research in recommender systems has been too focused on prediction accuracies of the underlying algorithms, assuming these accuracies were correlated to UX. They concluded, however, that there is no evidence for the presumed link of algorithm accuracy and UX. Instead, users' perceptions on interaction usability, appeal and perceived quality, affect UX much more.

The degree of transparency in the displays of recommender systems seems to be an important factor for user acceptance. Cramer et al. (2008) for example investigated the influence of transparency on user trust in, and acceptance of, content-based recommender systems in the cultural heritage domain and found that high transparency yielded higher acceptance, but not more trust. Different results in terms of trust were yielded by Mercado et al. (2016) who investigated the effect of transparency on trust and perceived usability using the example of an agent recommending plans for handling missions to the operator of an unmanned vehicle. Here, results indicated that trust and perceived usability increased as a function of transparency level.

During the above review of existing studies, no literature was found on in-car recommender systems for fatigue management. Therefore, it was worthwhile for this study to investigate the effect of different interaction strategies, as results from applications in other domains may not be transferable to driver fatigue management.

Summarizing, this study is expected to answer the following two RQs:

> RQ 4: Which effect do 4-minute thermal stimuli at 15 °C have on drivers' fatigue, when applied in response to the drivers' fatigue level?

> RQ 5: Which effect do different interaction strategies have on automation trust and acceptance of the fatigue management system?

This chapter elaborates on a study which applied both fatigue detection and fatigue intervention. First, the testing method is explained in Chapter 7.1. Second, the results in terms of stimulation effectiveness and preferred interaction strategies are reported (Chapter 7.2). Subsequently, the results are discussed in Chapter 7.3 and last, the conclusions are drawn in Chapter 7.4.

7.1 Method

This chapter deals with the method that was applied to address the above RQs. The subsequent chapters are structured as follows. Chapter 7.1.1 gives an overview of the technical setup in which the study took place. The study design is explained in Chapter 7.1.2 and the sample set is described in Chapter 7.1.3. The dependent variables collected in the study are listed in Chapter 7.1.4. Last, Chapter 7.1.5 outlines how the collected data were analyzed.

7.1.1 Setup

For this study, the BMW i3 apparatus, described in Chapter 3.3, was used in the flat-screen simulator (see Figure 7, Chapter 3.2). Cold air was generated with the car's AC unit and the thermal stimulus could be triggered remotely by the investigator and by the output of the simple logistic regression model for detecting fatigue which was introduced in Chapter 6.

7.1.2 Study Design

In this study, a fatigue management system was tested which repeatedly blew cold air at the driver when fatigue was detected. The cooling stimulus consisted of 15 °C cold air (COOL), which was circulated towards the driver's upper body for 4 minutes. This setting was found to be superior to the other tested settings in this thesis (Chapter 5.4). The rest of the time, the temperature was set to a thermo-neutral temperature of 24 °C. In order to avoid thermal discomfort of the driver, the minimum period between two cooling events was 2 minutes (see Figure 37). The earliest cooling was triggered after 6 minutes of driving. This is because the first minutes were necessary to record an individual baseline for each driver, which was a required input for the HRV calculation for the classifier. The total length of the drive was 34 minutes. The driving task was a monotonous highway drive with no traffic and predictable road environment. The previous study in Chapter 4 has shown that this type of simulated drive induces subjective and physiological fatigue after as little as 17 minutes of driving.

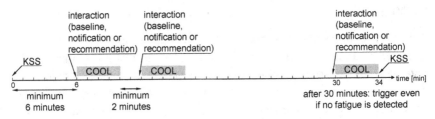

Figure 37:   Exemplary drive with three cooling events
Source: *Own illustration*

The fatigue detection method described in Chapter 6 was used in this study, which is based on HRV, pupil diameters and SCL measurements. The cooling was triggered when the simple logistic regression classified the driver as "fatigued", as long as the last cooling event ended more than 2 minutes ago.

The study employed a between-subject design with three groups. Each group was presented one of three interaction strategies for the cooling during the drive – baseline, notification and recommendation. The chosen strategies were the results of brainstormings with UX specialists of the BMW Group, pre-tests with small samples and several design iterations. The final strategies varied in terms of the degree of both transparency and automation of the system. All communication in the form of announcements, text displays and drivers' responses was in German and is described below.

Baseline group (n=31) – In this condition, the participants were neither informed nor provided any rationale regarding the cooling treatment.

Notification group (n=30) – In this condition, the driver was informed that cooling would be started. In detail, there was an audible earcon, followed by the announcement "I will refresh you" (translated from German by the author). At the same time, the car's central information display (CID) showed a sitting driver icon shown in Figure 38. Above the icon, it read "Refreshment" and below "4 minutes cooling" (translated from German by the author). This visualization was displayed for 15 seconds. After it disappeared, the cooling started.

Recommendation group (n=33) – Here, the driver was informed that fatigue was detected and asked whether he or she would like some refreshment. There was again an earcon, followed by the announcement "I think that you are getting tired. Would you like some refreshment?" (translated from German by the author). The visuals of the CID are shown in Figure 39. The display included the headlines "some refreshment?", "4 minutes cooling" and "fatigue detected" as well as the answer possibilities "yes, please!" and "no, thank you!" (translated from German by the author). The driver responded to this with a voice command. The participant's selection was highlighted in the CID, whereas the rest of the dialog disappeared. After a total display duration of 15 seconds the visualization ended and cooling started in case of a positive response.

Figure 38:   Notification display (simplified)   Figure 39:   Recommendation display
Source: Own illustration   (simplified)
                                                  Source: Own illustration

To ensure that all participants would experience the cooling (or the recommendation) at least one time per drive, a time-triggered start was enabled after 30 minutes just if the fatigue detector had classified the driver as "alert" throughout the earlier driving time (see Figure 37).

The study began with a 5-minute familiarization drive on a highway in which participants could get used to the simulation environment. During this phase, the notification or recommendation of the start of a seat massage was presented to the participants of the notification or recommendation group, respectively. This was done to train the participants to interact with the system without revealing that cooling would be applied during the experiment.

7.1.3 Participants

Ninety-eight volunteers participated in the study in March 2017. Details about the requirements for participation are described in Chapter 3.5. Four participants had to be excluded because of simulator sickness, a technical defect in the cooling system and two technical defects in the CID. Table 21 shows a description of the participants of the three groups.

There were no significant differences between the groups in terms of age, sleep duration or initial fatigue ratings which were provided before the familiarization drive. Differences in sleep duration and initial fatigue between the groups would have skewed parts of the subsequent analysis, because the KSS ratings recorded at the end of the drive would not only have been possibly affected by the number of cooling events and the interaction strategy, but also by a higher starting level fatigue in one group or circadian effects.

Table 21:   Description of samples for each group
Source: *Own table*

Parameter	Baseline	Notification	Recommendation
sample size	31	30	33
male/females	18 ♂, 13 ♀	21 ♂, 9 ♀	24 ♂, 9 ♀
age [y] (M, SD)	30.0 ± 9.3	31.4 ± 10.1	31.8 ± 11.0
sleep duration [h] in the night before the experiment (M, SD)	6.8 ± 0.8	7.0 ± 0.7	6.8 ± 1.0
initial KSS rating (M, SD)	4.3 ± 1.5	3.9 ± 1.4	4.1 ± 1.6

7.1.4 Dependent Variables

The participants provided a subjective estimation of their current fatigue level using the KSS (Åkerstedt and Gillberg, 1990; translated into German by Niederl (2007)) before the start of the experimental drive and at the end of it (Figure 37). In contrast to the studies in Chapter 5, the KSS was not asked during the drive because it would have affected the output of the regression model based on ECG, SCL and pupil diameter data. As the analysis of, e.g., HR (Figure 14, p. 75), SCL (Figure 16, p. 77) and pupil diameter (Figure 17, p. 77) showed, the verbal assessment of fatigue activated the drivers physiologically and would underpredict fatigue (Figure 36, p. 131). Instead of assessing fatigue verbally, only the objective fatigue assessment by means of the regression model was recorded during the drive.

At the end of the drive participants filled out questionnaires on automation trust with the 12-item unidimensional scale of Jian et al. (2000) (translated into German by Beggiato (2015)). To assess acceptance, the 5-point bipolar rating scale developed by Van der Laan et al. (1997) (translated into

German by Beggiato (2015)) was used. The drivers were also asked at the end of the study whether they a) liked the cooling and b) whether they perceived it as effective to mitigate fatigue as was asked in the previous studies (A3, p. 190). To compare the UX of the notification and the recommendation display, participants of these two groups also completed the user experience questionnaire (UEQ) (Laugwitz et al., 2008).

Details about the recording method of the physiological data are described in Chapter 3.6.

7.1.5 Data Analysis

The physiological measures, HRV, SCL and pupil diameters were processed in real-time in a Matlab Simulink model (A4, p. 199), which employed the regression model for fatigue assessment detailed in Chapter 6. The model consisted of a simple logistic regression model, which classified the driver as "alert" or "fatigued" and was used to trigger the interaction and the cooling for the three groups. Furthermore, a linear regression model was introduced in Chapter 6, which detected slight changes in fatigue, caused by sensory stimulation. The continuous fatigue estimation of the linear regression was used for post-hoc data analysis. Due to insufficient tracking quality of the pupil diameters, three physiological data sets from the recommendation group had to be excluded from the analysis.

For the evaluation of the cooling treatment effect duration, the means of eight non-overlapping 30-sec. intervals were analyzed. In order to determine how quickly the participants' fatigue levels returned to the pre-intervention level of fatigue, the fatigue levels during these 30-sec. intervals were compared to the fatigue level during the 30-sec. interval before the start of cooling.

7.2 Results

The following sections report the results in terms of system effectiveness (Chapter 7.2.1) and preferred interaction strategies (Chapter 7.2.2).

7.2.1 Results of the Effectiveness of Fatigue-Based Cooling

This study provided the opportunity to evaluate the accuracy of the fatigue classification with online data. For the evaluation of accuracy, the subjective fatigue ratings were used as ground truth. The KSS ratings provided before the start of the drive were compared with the output of the simple logistic regression at the beginning of the drive. The KSS ratings provided at the end of the drive were compared with the output of the simple logistic regression in the last minute of the drive. A classification was considered correct, when the KSS value ranged between 1 and 4 and the simple logistic regression output "alert", as well as, when the KSS value ranged between 8 and 9 and the simple logistic regression output "fatigued". This resulted in a classification accuracy of 75 % and the confusion matrix of the results is shown in Table 22.

Table 22: Confusion matrix for fatigue estimation
(digits represent the absolute amount of observations)
Source: Own table

Classified as →	Alert	Fatigued
KSS 1, 2, 3, 4	53	14
KSS 8, 9	12	25

There was a false positive rate of 13 %, which means that false fatigue alarms were issued when the driver was not actually fatigued according to the KSS data. There was a false negative rate of 12 %, which are missings in the fatigue detection. All observations with intermediate KSS value were not considered in the evaluation of accuracy, as the simple logistic regression is not suited for these values (see Chapter 6).

To assess the relationship between the continuous outputs of the linear regression model for fatigue and the KSS ratings, a correlation was performed. This yielded a correlation coefficient of $r=0.50$, $p<.001$. Figure 40 shows the development of fatigue over the course of the drives according to the linear regression model.

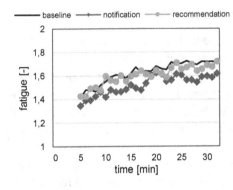

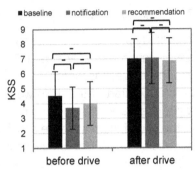

Figure 40: Mean fatigue prediction over the course of the drive for each group

Source: Own illustration

Figure 41: KSS ratings (M and SD) before and after the drive for each group

(- not significant,
* significant (p<.05))

Source: Own illustration

A mixed-design ANOVA was conducted that examined the effect of time-on-task (beginning and end of drive) and interaction strategy on the predicted fatigue level. There were no statistically significant interactions between the effects of time-on-task and interaction strategy, $F(2, 88)<1$, $p=.89$. Simple main effects analysis showed that the fatigue levels at the end of the drive were significantly higher than at the beginning ($F(1, 88)=83.8$, $p<.001$), but there were no differences between the interaction strategies ($F(2, 88)=1.3$, $p=.27$).

This aligns with the analysis of the KSS ratings (see Figure 41), for which there were also no statistically significant interactions between the effects of time-on-task and interaction strategy, $F(2, 91)=1.97$, $p=.15$. The KSS ratings after the drive were significantly higher than before the drive ($F(1, 91)=282.6$, $p<.001$), and there were no differences between the interaction strategies ($F(2, 91)<1$, $p=.44$).

Table 23 shows the number of cooling offers per group and the number of actual coolings. A Kruskal-Wallis test showed that there was no significant difference in the number of offers between the different groups ($H(2)=3.3$, p=.19). The recommendation group, however, declined the cooling offer in 38.1 % of the cases, which is why there was a significant difference in the number of actual coolings (H(2)=12.3, p<.05). Post-hoc tests with Bonferroni correction showed that the recommendation group experienced significantly less coolings than the baseline group (U=261, z=-3.4, p<.0167). Two of the participants in the recommendation group did not experience cooling at all, because each rejected the single time-triggered offer, they had received after 30 minutes. In contrast, there were nine drivers in this group that accepted every single offer.

Table 23:   Amount of coolings and impressions of each group
Source: Own table

	Baseline	Notification	Recommendation
number of cooling offers (Mdn) (p=.19)	4	3	4
number of actual coolings (Mdn) (p<.05)	4	3	2
Did you like the cooling? (p=.12)	53.4 % yes	70.0 % yes	77.4 % yes
Do you think that the cooling helps against fatigue? (p=.998)	87.1 % yes	86.7 % yes	87.1 % yes

The percentage of drivers who liked the cooling was the highest in the recommendation group (77.4 %) and the lowest in the baseline group (53.4 %) (see Table 23). The difference of these positive-feedback rates, however, is on a non-significant level (χ^2(2, N=91)=4.18, p=.12). The perceptions of the participants on the effectiveness of cooling against fatigue was similarly high in all groups (ca. 87 %, χ^2(2, N=91)=0.003, p=.998) (Table 23).

Figure 42 shows the averaged fatigue levels during cooling in the baseline group. The data of one participant who experienced the compulsory start of the cooling at the end of the drive was excluded from the analysis. When the drivers experienced the cooling for the first and second time, fatigue stayed significantly lower for the next 3.5 minutes of cooling as t-tests showed (Table 24), compared to the peak at the beginning of cooling. The third cooling did not yield any significant reduction in fatigue.

Similar results were seen in the notification group (see Figure 43). For the first cooling, fatigue stayed significantly lower for the entire 4 minutes of cooling, compared to the peak 15 seconds before the beginning of cooling, when the notification appeared on the CID. When these drivers experienced the cooling for the second time, fatigue was significantly reduced for only 2.5 minutes – after this period, the difference in fatigue was not significant (Table 25). The third cooling did not yield any significant reduction in fatigue.

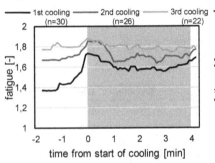 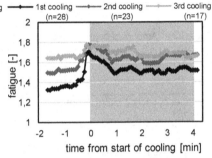

Figure 42:	Mean fatigue prediction before and during cooling for the baseline group
Source:	Own illustration

Figure 43:	Mean fatigue prediction before and during cooling for the notification group
Source:	Own illustration

Table 24:	T-scores for the comparison of 30-second intervals for the baseline group
	(reference value is the fatigue level at the beginning of cooling, * significant (p<.05))
Source:	Own table

Table 25:	T-scores for the comparison of 30-second intervals for the notification group
	(reference value is the fatigue level at the beginning of cooling, * significant (p<.05))
Source:	Own table

Period after start of cooling [s]	0-30	30-60	60-90	90-120	120-150	150-180	180-210	210-240
1st cooling	2.1*	3.0*	2.2*	2.8*	2.8*	2.3*	2.1*	1.0
2nd cooling	3.7*	4.5*	3.2*	3.5*	3.7*	3.2*	3.1*	1.6
3rd cooling	1.2	1.4	1.3	1.5	1.8	1.8	1.4	1.8

Period after start of cooling [s]	0-30	30-90	60-90	90-120	20-150	150-180	180-210	210-240
1st cooling	2.0	2.4*	2.3*	2.7*	2.5*	2.7*	2.2*	2.8*
2nd cooling	0.9	2.4*	1.4	1.9	2.6*	1.7	1.1	1.1
3rd cooling	0.3	1.2	0.6	1.7	1.7	1.1	1.1	1.4

Figure 44 shows the fatigue levels for the recommendation group, in the case where they accepted the offer. Again, the first cooling reduced fatigue significantly for the entire 4 minutes of cooling, whereas the effect of the second cooling only lasted for 2 minutes (Table 26). For this group, there was also a significant reduction of fatigue during minute 2 and 3 of the third cooling.

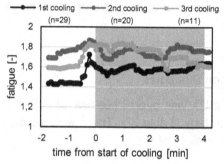

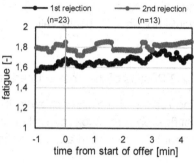

Figure 44:   Mean fatigue prediction before and during cooling for the recommendation group

Source: Own illustration

Figure 45:   Mean fatigue prediction before and after the offers in case those were rejected

Source: Own illustration

Table 26:   T-scores for the comparison of 30-second intervals for the recommendation group

(reference value is the fatigue level at the beginning of cooling, * significant (p<.05))

Source: Own table

Period after start of cooling [s]	0-30	30-60	60-90	90-120	120-150	150-180	180-210	210-240
1st cooling	4.0*	2.7*	3.3*	2.6*	3.1*	2.7*	2.5*	2.1*
2nd cooling	2.4*	1.9	2.7*	2.5*	1.0	0.8	1.2	1.1
3rd cooling	0.8	0.9	1.0	1.9	2.7*	3.0*	2.2	1.4

The results of the occasions when the drivers of the recommendation group rejected the offers are shown in Figure 45. In this case, there were no differences in fatigue between the time of the offer and any of the subsequent minutes.

The results of the fourth and fifth cooling as well as the third, fourth and fifth rejection are not presented because only few drivers experienced cooling or rejected cooling this often. Therefore, an analysis of these times would not be statistically valid.

7.2.2 Results of the Tested Interaction Strategies

The results of the UEQ showed that UX was overall positive for both the notification and recommendation strategy (Table 27). There were no significant differences in any of the dimensions of UX.

Table 27:   User experience dimensions (M and SD)
Source: *Own table*

	Notification	Recommendation	T-test result
attractiveness (M, SD)	1.4 ± 1.0	1.2 ± 0.9	t=0.7, p=.48
perspicuity (M, SD)	2.2 ± 0.6	1.9 ± 0.8	t=1.6, p=.11
efficiency (M, SD)	1.6 ± 0.7	1.3 ± 0.6	t=1.9, p=.07
stimulation (M, SD)	0.8 ± 1.1	0.7 ± 0.8	t=0.3, p=.76
novelty (M, SD)	0.4 ± 1.5	0.1 ± 1.3	t=0.8, p=.41

The results of the automation trust questionnaire are shown in Figure 46. The different interaction strategies had a significant effect on the trust ratings of the fatigue management system ($F(2, 91)=3.59$, $p=.03$). The significant difference was between the baseline and the recommendation strategy. The difference in the mistrust between the different concepts was on a non-significant level ($F(2, 91)=2.87$, $p=.06$).

In terms of the acceptance dimensions, there was neither a significant difference in usefulness (F(2, 91)=1.56, p=.22) nor in satisfaction (F(2, 91)=2.07, p=.13) (Figure 47).

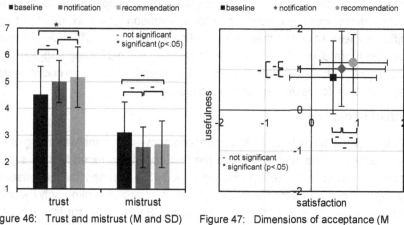

Figure 46: Trust and mistrust (M and SD)   Figure 47: Dimensions of acceptance (M
            in automation   and SD)

Source: Own illustration   *Source: Own illustration*

7.3 Discussion

This study provided an opportunity to evaluate the accuracy of the fatigue regression model in a closed-loop because both predicted and subjective fatigue were recorded, allowing the comparison of these two. The 75 % accuracy of the classifier is fair, considering that the participants were not sleep-deprived (see discussion in Chapter 6.3). Furthermore, the output of the regression model correlated well (r=0.50) with the subjective fatigue ratings. This correlation coefficient is higher than the ones yielded with single physiological parameters in Chapter 4.2.3 and 5.1.2.3 which were around |r|=0.3. The results also show that the fatigue prediction reflected slight changes in fatigue as could be seen in Figure 42, Figure 43 and Figure 44 and therefore allowed for successful significance testing.

Next, the monotonous drive in this study successfully induced fatigue, as a main effect of time was observed in both the predicted and the subjective fatigue ratings. For this reason, the cooling stimulus was successfully triggered during the drives. Because 38 % of the coolings in the recommendation group were rejected, the recommendation group experienced significantly less coolings than the baseline group. The 15°C cold air likely caused thermal discomfort and led to rejections. This finding suggests further that the chosen minimum period of 2 minutes in between coolings was not sufficient for restoring thermal comfort.

Effect of cooling

The results of the effectiveness of cooling showed that the wakening effect was subject to habituation: While the first cooling reduced fatigue significantly for almost the entire cooling period (3.5 minutes for the baseline group and 4 minutes for the other groups), the second cooling only reduced fatigue for about half of the cooling period (3.5 minutes for the baseline, 2.5 minutes for the notification- and 2 minutes for the recommendation group). The third cooling did not show an effect in the baseline and notification group. In the recommendation group the cooling effect only lasted for one minute (Table 26).

This showed, that the participants got used to the stimulus upon repetition. This was expected, as studies with repeated stimuli showed that immediate physiological responses like pupillary dilatation and HR reactions would be less apparent upon repetition (Goldwater, 1972; Graham and Clifton, 1996). Therefore, the regression model for fatigue assessment, which uses these physiological signals as inputs, also detected this habituation effect.

The fourth and fifth cooling were not included in the significance testing, as the sample of drivers experiencing cooling this often in the study, was too small for the results to be meaningful. However, as habituation effects were observed in the third cooling, it is likely that upon fourth and fifth repetition, no effects would be seen.

These results can be summarized to answer RQ 4:

> A 4: The first 4-minute thermal stimulus of 15 °C reduces fatigue for approximately 4 minutes, whereas the second stimulus only reduces fatigue for approximately 2.5 minutes. The third cooling is hardly effective and does not yield consistent significant fatigue reductions.

One limitation of the study in terms of testing for fatigue reduction was that there was no control condition without cooling and hence, the point in time before cooling was used as the control minute for the significance testing. This skewed the results because fatigue at different points in time were used for the comparison and it is known from earlier studies, that time has an effect on fatigue. Proper control conditions would probably have yielded slightly higher fatigue ratings, and consequently resulted in more significant test results.

As the ANOVA results of the predicted and the subjective fatigue showed, the different interaction strategies did not have an effect on fatigue. The fact that the recommendation group did not feel more fatigued than the other groups, seems counter-intuitive, as they experienced significantly less cooling events than the other groups (Table 23). This can be answered though, by the short-lived effect of cooling. This means, that by the end of the drive, the participants in all groups have redeveloped their physiological and subjective fatigue. This result is particularly interesting, with regard to placebo effects, as the announcement ("I think that you are getting tired.") may have led to participants in the recommendation group feeling more tired based on the auditory suggestion. Looking at the non-significant differences in KSS ratings, however, it can be seen that there was no change in perceived fatigue because of any placebo effect.

When comparing the positive-feedback rates of the baseline group (Table 23) with the results of Chapter 5.2, one might ask why the percentage of people liking the cooling, is worse in this study (53 % compared to 85 %), as the same stimulus was applied. The reason for this discrepancy is the larger number of cooling events, participants experienced in this study: The positive-feedback rates of drivers, who experienced cooling four or five times during the drive is very low and thus diminishes the mean of all drivers. This is another reason, along with the inefficiency of repeated cooling due to habituation, that extensive application of 15 °C cold air on the driver's upper body should be avoided.

Effect of interaction strategy

Next, the results of the different interaction strategies are discussed. First, it is important to highlight, that there was no difference between the notification and the recommendation display in terms of UX (Table 27). This means, that both concepts were equally attractive, novel, stimulating, efficient and, most importantly, perspicuous for the drivers. This is essential because any of the reported effects on acceptance or trust were not subject to any side effects of the display design.

The results of the automation trust scale showed that the recommendation strategy was superior compared to the baseline strategy with no communication (Figure 46). This result aligns with the studies of Mercado et al. (2016) and Sinha and Swearingen (2002) who also found increasing trust ratings when recommendations for handling missions and music were more transparent. The participants of the recommendation group revealed in follow-up questions after the drive, that they thought positively about the fact that they were asked whether cooling should start. The people of the baseline group, on the contrary, felt undecided about the fact, that cooling started automatically. Moreover, all concepts yielded positive trust ratings, with a minimum average of 4.5 and an absolute difference of about 0.6 between the highest and the lowest trust ratings on a scale from 1 to 7.

The results of the acceptance scale showed that the notification and recommendation group were rated as more useful and satisfying than the baseline group, however, both differences were on a non-significant level (Figure 47). This aligns with the also non-significantly different positive-feedback rates of the cooling (see Table 23). The fact, that usefulness was not affected (Figure 47), is also in alignment with the results of Table 23, which shows that the perceived-effect rates were more or less the same in all three groups.

Understanding these results helps answer RQ 5:

> A 5: The interaction strategy has an effect on users' automation trust of the system. The strategy to recommend cooling as a treatment against the detected fatigue, is perceived as more trustworthy than an automated start of cooling without the communication of what will happen and why. The interaction strategy does not have an effect on the acceptance of the fatigue management systems of drivers.

One limitation of the design of the study was that the different interaction strategies varied in terms of the degree of both transparency and automation of the system. This did not allow for a direct inference of the effects on trust and acceptance from changes in transparency or automation. A proper isolation of the independent variables would have allowed for more general conclusions on the effect of transparency and automation, instead of the effect of specific interaction strategies.

7.4 Conclusion

The above study tested a closed-loop of fatigue detection and fatigue intervention. This allowed for evaluating the effectiveness of thermal stimulation when applied upon fatigue detection, instead of applying the stimulation after a pre-defined amount of time. Furthermore, the study analyzed which interaction strategy for announcing the thermal stimulation yields higher trust and acceptance of the driver.

First, the results of the repetition of cooling showed that only the first and second cooling are effective against fatigue from a physiological point of view. This means that future systems can reduce passive fatigue only in the short term and, thus, it is worthwhile to examine stronger stimuli or variations of multi-sensory stimulation that may be less subject to habituation. According to Parasuraman et al. (1998) a fatigue management system should include a variety of different stimuli in the vehicle interior in order to increase the driver's vigilance.

Second, the results of the interaction strategy have shown superiority of the recommendation option for several reasons. The first reason would be, that drivers made use of the "reject" option and this result on its own, is reason to avoid the automated start of the coolings. Next, cooling tended to be more likeable among the drivers who interacted with the recommendation system. And last, the drivers' automation trust in the fatigue management system was higher in the group that was provided with the reason for the cooling recommendation, than in the group which did not see any reasoning or notification before the cooling.

8 Discussion

Chapter 8.1 recaps the RQs of this thesis and the studies' contribution in the field of driver fatigue. Chapter 8.2 deals with the overall methodical and applicational limitations of the thesis. The chapter closes with an overall discussion of the studies in Chapter 8.3.

8.1 Summary

The focus of this thesis was to empirically investigate the effect of thermal stimulation as a countermeasure against passive fatigue in human subjects while driving. After a literature-based analysis of the research gap in Chapter 2 and the selection of methods in Chapter 3, a pre-test was conducted to assess the feasibility of inducting passive fatigue in study participants using monotonous drives, thus providing an answer to the first RQ:

> RQ 1: Which degree of passive fatigue can be induced by means of task underload in drivers according to subjective and objective measures?

The first study on fatigue induction showed that monotony caused by simulated driving on a highway with very little traffic evoked, on average, subjective fatigue ratings describable as foggy or feeling slowed down after as little as 17 minutes of driving (Chapter 4). After this time, subjective and physiological indicators of fatigue, such as HRV, were significantly higher compared to more challenging driving scenarios. This result was used as a reference for the design of the simulator based driving studies for the investigation of thermal stimulation as a countermeasure against passive fatigue, which helped to answer the second RQ:

> RQ 2: Which effects do short-term thermal stimuli have on the drivers' passive fatigue?

© Springer Fachmedien Wiesbaden GmbH, part of Springer Nature 2020
E. Schmidt, *Effects of Thermal Stimulation during Passive Driver Fatigue*,
Gestaltung hybrider Mensch-Maschine-Systeme/Designing Hybrid Societies,
https://doi.org/10.1007/978-3-658-28158-8_8

The main studies of this thesis, which address RQ 2, consisted of three driving simulator studies, each investigating a different setting of the vehicle's AC. To explore each setting, studies were conducted comparing both subjective and physiological fatigue measures during thermal stimulation to a thermo-neutral control condition.

In the study (Chapter 5.1), which applied a 6-minute thermal stimulus of 17 °C cold air along with high fanning intensity towards the upper body, an SNS activation was measured via the reactions of the skin conductance and pupil diameter, which increased significantly (Figure 16 and Figure 17) during the cooling. Another indication of reduced fatigue were fewer observations of eye closures (Figure 18). In addition, the subjective fatigue ratings were significantly lower after the thermal stimulation (Figure 13). However, only 59 % of the participants indicated to like this treatment.

The next study (Chapter 5.2) compared the effectiveness of the application of 15 °C cold air for different durations, namely 2 minutes and 4 minutes. The participants reported significantly lower subjective fatigue ratings (Figure 21a and b) for both treatments. However, only the 4-minute stimulus yielded bradycardia and an increase in pupil diameter (Figure 24 and Figure 25) whereas the 2-minute stimulus did not cause pupillary dilatation (Figure 23). For this reason, the 4-minute stimulus was superior to the 2-minute stimulus in terms of effectiveness against passive fatigue. A portion of 85 % of drivers liked the thermal stimulus.

In Chapter 5.3, the study results of lower leg cooling with 15 °C cold air showed that this treatment was not effective against passive fatigue. Although participants perceived significantly reduced fatigue according to the KSS-values (Table 14), the physiological measures HR (Figure 28), SCL (Figure 30) and pupil diameter (Figure 31) did not confirm the hypothesis that leg cooling would reduce fatigue. Because physiological indicators did not change in response to the treatment at any point in time, the study could not confirm fatigue mitigation, since the subjective ratings may have been the results of a placebo effect. The percentage of drivers liking the stimulus on the legs was also low with 62 % (Table 16).

A systematic comparison of the three studies in terms of effectiveness and comfort was presented in Chapter 5.4. First, the comparison showed that both studies on upper body cooling (Chapter 5.1 and 5.2) revealed physiological and subjective indications of reduced fatigue, while the study on leg cooling (Chapter 5.3) failed to cause effects in physiological measures. Second, the positive-feedback rates showed superiority of the 2-minitue and 4-mintue upper body cooling. Because of the superiority of the 4-minute upper body cooling at 15 °C in terms of comfort and its effect on KSS, HR and pupil, this setting was recommended for further studies.

The next RQ dealt with the objective measurement of fatigue which is essential to determine the point in time when thermal stimulation should be applied and for the continuous evaluation of fatigue during the application of a countermeasure:

> RQ 3: At which accuracy can passive fatigue be detected based on physiological measures?

In order to investigate how accurately physiological indicators can predict fatigue, a secondary data analysis of a series of simulator studies which provided the training data was performed (Chapter 6). The results of the fatigue model based on the drivers' ECG, skin conductance and pupil diameter measurements showed that passive fatigue of non-sleep-deprived drivers could be classified with an accuracy of 77 % using these input signals in a simple logistic regression. The test and validation of a linear regression using the same inputs further demonstrated the ability to model slight decreases in fatigue caused by thermal stimulation.

Having created a knowledge base of the above three RQs, it was possible to implement the findings in a final study. This study incorporated the online fatigue assessments based on physiological data and the 4-minute thermal stimulus, which was triggered when fatigue was detected. This study aimed to provide insights in order to answer the fourth RQ:

> RQ 4: Which effect do 4-minute thermal stimuli at 15 °C have on drivers' fatigue, when applied in response to the drivers' fatigue level?

The results showed that the initial cooling application reduced physiological fatigue significantly for the entire cooling period (Figure 42, Figure 43 and Figure 44). The second cooling, in contrast, reduced fatigue on a significant level only for about 2.5 minutes, while the third cooling was not effective as the change in fatigue was not significant (Chapter 7).

To complete the research on a fatigue management system, which detects and counteracts fatigue, a question was directed towards the future HMI with such a system:

> RQ 5: Which effect do different interaction strategies have on automation trust and acceptance of the fatigue management system?

The study on automation trust of different interaction strategies showed that a fatigue management system that recommended cooling as a treatment against the detected fatigue, was perceived as more trustworthy than an automated start of cooling without communication. The interaction strategy did not have an effect on the acceptance of the fatigue management systems of drivers (Chapter 7).

Contributions

The studies of this thesis contribute to a better understanding of fatigue management by means of thermal stimulation in the field of driver fatigue. Insights provided by this thesis are a **scientific evaluation of the effect of different thermal stimuli** during monotonous driving on both subjective and objective fatigue indicators. The results showed – unlike other studies testing weaker stimulation on non-sleep-deprived drivers (see Table 6) – that **thermal stimulation of the upper body of a driver can reduce passive fatigue**. The studies furthermore showed that the **physiological activation is short-lived and subject to habituation**.

Another result of this thesis is the **model for detecting and quantifying passive fatigue** by means of a logistic and linear regression model, respectively. Compared to existing models (see Table 3), the regression model stands out because it focused on passive fatigue and was trained and cross-validated with data from a large amount of drivers (n=88) who were not sleep-deprived.

Other insights provided by this thesis, relate to the **HMI strategy of a fatigue management system**. The results showed that drivers had more trust in the system when the cooling was recommended along with an explanation that fatigue was detected, instead of a fully automated system.

8.2 Methodical and Applicational Limitations

The studies' limitations in terms of method and future application are discussed in the following.

Methodical limitations

The first methodical limitation of the studies was that all studies were **performed in static driving simulators**. Although the lack of visuomotor stimulation was contributing to the aim of fast fatigue induction, it did not reflect real driving conditions in which fatigue needs more time to develop (Philip et al., 2005; Hallvig et al., 2013; Fors et al., 2016). Real driving studies investigating the fatigue countermeasures would require the possibility of the examiner to take over the driving task for ethical reasons, since the participants become increasingly impaired due to fatigue and are at risk of causing an accident while driving on the road. Furthermore, several examiners would be required per study participant in order to allow them to work in shifts. This is necessary, as examiners would become increasingly fatigued during real driving, as well.

Investigating the effects in real driving conditions would be particularly interesting because the effect of thermal stimulation may be increased since the drivers would be less fatigued based on the visuomotor stimulation and the cooling might relieve the strain from drivers more efficiently. Conversely, the effect of thermal stimulation could also be decreased, as it may be less noticeable in combination with the visuomotor stimulation according to the resource theory (Helton and Russell, 2012).

The second limitation was that a **non-random sample set of BMW Group employees was recruited** for the studies. As described in Chapter 3.5 the convenience sample differed from the general population in terms of nationality, gender, age and also health because the participants confirmed at the beginning of each study that they feel healthy and fit enough to participate in the study. Even though not explicitly recorded, the sample probably also differed to a general population in terms of brand loyalty, since previous findings suggest the existence of employees' loyalty to the brand by which they are employed (Fram and McCarthy, 2003; Locascio et al., 2016). To what extent these factors may cause a bias in the measured effect of thermal stimulation on fatigue is discussed in the following.

An extensive literature review by Parsons (2002) addressed the thermal comfort in special populations. His review included the effects of geographic location, gender and age on thermal comfort. He summarized that despite all notions of, e.g., females and elderly preferring warmer environments or people from hot climates requiring different thermal conditions for comfort than people from cold climates, the geographic location, gender and age do not affect thermal comfort perceptions. Notions as above can be attributed to differences in metabolic rate or clothing which are the only two human factors that influence the thermal comfort (see Chapter 2.2.1). Therefore, it was concluded that the thermal comfort ratings of the BMW Group employees should not differ from ratings provided by a general population.

Even though thermal comfort perceptions of the sample were reasoned not to be biased, the physiological reactions upon the provided stimuli probably cannot be generalized to a general population. This is because the studies of LeBlanc et al. (1975, 1978) have shown, that facial cooling effects on physiological measures differed between people of different ages, physical fitness and temperature acclimatization. Taking these results into account, the physiological reactions of the BMW Group employees can be assumed to have been more prominent compared to a general population because of their younger age and higher fitness.

The above mentioned brand loyalty of employees plays a role in the validity of usability tests. Locascio et al. (2016) showed that employees differ from a general population when it comes to rating competitor's products, but not for rating the company's own products. Therefore, they suggested that using employees for internal usability testing of the company's own products may be valid. Concluding from these findings, it can be assumed that the BMW Group employees did not rate the thermal stimuli as more positive than a general population because of their loyalty to the brand.

Summarizing, the method of convenience sampling applied for the studies of this thesis, caused a distortion in the physiological results based on the sample's young age and the associated sensitive autonomic response. No substantiation was found for assuming that the subjective responses about the effect of thermal stimulation was confounded by the sample's nationality, age and gender distribution or their loyalty to the brand.

Another methodical limitation was that the employed **study design changed** halfway through the thesis. Whereas the studies in Chapter 4, 5.1 and 5.2 were carried out as a within-subject design, the studies in Chapter 5.3 and 7 employed a between-subject design. The reason for starting out with within-subject designs was that this method automatically controls for individual variability and requires less participants than between-subject testing (Nielsen, 1994). Though, the within-subject tests of the studies lasted up to two hours and it became clear that it was hard to recruit participants for this duration because of their restrictions during work hours, i.e., limited time in between meetings. The between-subject design made it easier for the employees to accommodate participation because a test usually lasted only up to one and a quarter hours then, resulting in a higher response rate. Therefore, the study design was changed to a between-subject design in the last two studies. The assignment of participants in groups in between-subject testing risks a bias (Nielsen, 1994, p.178) and therefore the assignment was done in a counterbalanced way.

The fourth methodical limitation was that **the remote eye trackers used did not output PERCLOS** as many head-mounted eye trackers do. This was a limitation, as several studies have shown that PERCLOS has a strong correlation with fatigue. Remote eye trackers were used in these studies to prevent participants' comfort from being restrained by wearing a head-mounted eye tracker. The PERCLOS value might have been a good input for the fatigue detection and quantification model of this thesis as it might have increased the detection accuracy. However, PERCLOS might have been confounded at the times of thermal stimulation, because it has been shown, that the thermal environments – especially air movement and humidity – affect blink rates (Acosta et al., 1999; Tsutsumi et al., 2007).

The next methodical limitation was that the participants did **not start the experimental drives with an exact equal initial KSS rating and the amount of sleep in the night before the experiment varied**. Furthermore, the participants partly drove during their circadian afternoon dip, which is why a minor influence of SR fatigue could not be ruled out.

Applicational limitations

The results of this thesis are also limited in terms of their future application. First, the suggested countermeasure against fatigue is **only effective against passive fatigue**. Passive fatigue occurs in monotonous settings due to lack of stimulation. Therefore, additional stimuli, i.e., mental, haptic, thermal and olfactory mitigate this type of fatigue (see Figure 3, p. 15). Thermal stimulation will likely not be effective against active fatigue, which is caused by task overload, and SR fatigue (see Figure 2). This limitation is crucial, as it necessitates an accurate fatigue quantification, which can deduce the cause of fatigue in order to make use of the appropriate countermeasure.

Another application-related limitation is that cooling only **reduces fatigue in the short term**. Therefore, it can merely assist the drivers to reach the next parking lot or to take other measures instead of helping to continue the journey.

Finally, the proposed countermeasure against fatigue is **limited to vehicles with separate climate zones** for driver and passengers, in order to prevent discomfort for passengers during the system's intervention.

8.3 Discussion of Studies

The studies of this thesis on upper body cooling showed that thermal stimulation of the upper body caused significant changes in physiological parameters of a driver which could be associated with an activation of the SNS. This is because pupillary dilatation, the increase in skin conductance, as well as HR deceleration, are a **consequence of sympathetic activation** caused by sensory stimulation as previous studies outside the driving context have shown (Bradley et al., 2008; Goldwater, 1972). In contrast to the study on hand cooling of Van Veen (2016), significantly lower subjective fatigue ratings suggested successful fatigue mitigation via thermal stimulation. In absolute terms, fatigue was not eliminated, but rather slightly mitigated. Chapter 5.1 showed that the fatigue levels after 16 minutes of driving were similarly high as the fatigue ratings provided directly after cooling after 26 minutes which pointed to the limitations of thermal stimulation.

The **reduction of HR during upper body cooling** did not seem to align with the increased wakefulness of the driver at the first glance, however, this was a physiological response to the short-term thermal stimulus, which has been observed with facial fanning in non-vehicle settings (Collins et al., 1996; Hayward et al., 1976; LeBlanc et al., 1975, 1976, 1978). Researchers have suggested that the decrease of HR is caused by vagal reflex in which facial receptors initiate a stimulation of the trigeminal nerve (Collins et al., 1996; LeBlanc et al., 1976). The increase in HR through hand cooling observed by Van Veen (2016) is not contradictory to the results in this thesis on upper body cooling since it has been proven that hand and facial cooling cause different HR responses (LeBlanc et al., 1975, 1978).

In contrast to upper body cooling, **leg cooling did not yield physiological activation**, even though the same stimulus temperature and duration were successfully causing physiological activation in a previous study on upper body cooling. The non-significant results in the physiological measures could be for two reasons. First, there are fewer thermal cold receptors in the skin of the legs (see Table 4 in Chapter 2.2.1) and, second, the leg clothing of the participants could have insulated the legs too much. Therefore, thermal stimulation of the drivers' legs is not an ineffective countermeasure as long as specific criteria can be fulfilled. In order to achieve fatigue-mitigating effects by means of leg cooling, the temperature of the air that directly surrounds the legs has to be lower. This consideration can be substantiated with the study of Janský et al. (2003), who could show physiological effects, indicating an SNS activation, by means of leg cooling in a 12 °C water bath.

Another insight of this thesis to be discussed is the **short-lived effect of thermal simulation**. The non-significant difference in KSS ratings after the end of the COOL and the CONT drive in Chapter 5.2 (Figure 20) showed that 6 minutes after the thermal stimulus had ended in the COOL condition, the KSS ratings were at the same level as in the CONT condition. This aligned with the effects in HR and pupil diameter in the study of Chapter 5.2 which disappeared by the time the stimulus has ended. The short-lived effect could be reaffirmed by the last study in Chapter 7 because the continuous fatigue assessment based on physiological measures showed that the effect on physiology did not last longer than 4 minutes.

Furthermore, the study in Chapter 7 on repeated stimuli showed, that there was a **decrease of the activating effect of cooling upon repetition**. The decrease in the physiological effects was due to habituation, which aligned with other studies on repeated stimuli (Goldwater, 1972; Graham and Clifton, 1996) that have shown that immediate physiological responses like pupillary dilatation and HR reactions would be less apparent upon repetition outside of the driving context.

To this point, the thesis' results were discussed referring to studies on sensory stimulation mainly outside of the driving context. This is because there are few driving studies available that applied thermal stimulation and recorded the same physiological signals. At this point, however, it is worthwhile to also qualitatively compare the thesis's results on thermal stimulation with different types of stimulation applied while driving, such as mental, haptic and olfactory stimulation as presented in Figure 3.

Mental – Mental activity through a 20-minute interactive-cognitive-task (Trivia) has been shown to decrease subjective fatigue significantly by an average of 2 points on the 11-point Swedish Occupational Fatigue Inventory (Gershon et al., 2009a) and to even reduce HRV, which never occurred in the studies of this thesis. Also Jarosch et al. (2017) reported significantly lower levels of subjective fatigue and significantly lower PERCLOS while engaging in a quiz task compared to a monotonous task while driving in autonomous mode (level 3 according to the SAE Standard J3016). In their study, the KSS ratings after 19 minutes of performing the quiz task were on average 2 points lower than in the monotonous setting. The largest difference in KSS ratings achieved by thermal stimulation in this thesis was 1.3.

Haptic – Torque on a steering wheel for 3 km (at a speed of 40 km/h) during passive fatigue yielded significantly lower PERCLOS and lower values of LF/HF during assisted driving (level 1 according to the SAE Standard J3016) (Wang et al., 2017). Merat and Jamson (2013) showed that rumble strips on a 3-km-long stretch of street reduced PERCLOS, but pointed out that subjective fatigue ratings provided at the end of the drive suggested that the effect of the stimulus must have been short-lived, which is comparable to the results of thermal stimulation in this thesis.

Olfactory – The 30-minute simulator study of Mahachandra and Garnaby (2015) found that continuous exposure to peppermint did not yield significantly higher alertness than a placebo fragrance based on EEG measures, but they argued that a different exposure method might yield larger effects.

These studies on different types of stimulation should serve as a qualitative reference, however, a quantitative assessment in terms of effect sizes of different types of stimulation is not possible. This is because the studies of this thesis on thermal stimulation used for example different fatigue induction methods and durations, different driving simulators (fixed-base vs. moving-base), different stimulus durations and different vehicle operation modes (manual vs. automated). Therefore, no statement can be made about whether thermal stimulation is more or less effective against passive fatigue than other types of stimulation. A direct quantitative comparison of the effects of mental, haptic, thermal and olfactory stimulation would require equal study designs and procedures.

9 Conclusion and Outlook

In this chapter conclusions of the gained insights are drawn in terms of implications for practical applications (Chapter 9.1) and directions for future research (Chapter 9.2).

9.1 Implications for Practice

Today, 99 % of new cars have AC systems (Motor Trend Group, 2010), which is a solid starting point for the following discussion on the implications of the thesis' results because the availability of an AC is not a limiting factor in the implementation of an AC-assisted comfort system for mitigating fatigue.

More **limiting by far is the availability of reliable fatigue detection systems**. As reviewed in Chapter XXV, most of the available systems lack scientific validation of the detection accuracies and current technologies do not fulfill guidelines for a legally defensible system (Dawson et al., 2014; Schmidt, 2018). Furthermore, current systems are based on fatigue symptoms such as steering behavior (Volkswagen AG, 2018), lane keeping behavior (Volvo Cars, 2018) or gaze behavior (Toyota, 2018) and not based on the causal factors of fatigue, such as sleep deprivation, task overload or task underload (Figure 2). This would be crucial information for a fatigue detection device because it greatly impacts the selection of appropriate countermeasures (Van Veen et al., 2014).

A solution to the insufficient **contextual awareness** of the current system is offered by Fletcher et al. (2005), who introduced a road scene monotony detector which is based on video streams of the frontal driving scene and found a high correlation with human ratings of monotony. Several technologies are being developed with the potential to support the accurate detection of fatigue. The first is the rapid progress in sensors that allows, e.g., for measuring people's movements, behavior, physiological data and sleep

© Springer Fachmedien Wiesbaden GmbH, part of Springer Nature 2020
E. Schmidt, *Effects of Thermal Stimulation during Passive Driver Fatigue*,
Gestaltung hybrider Mensch-Maschine-Systeme/Designing Hybrid Societies,
https://doi.org/10.1007/978-3-658-28158-8_9

cycles with devices such as smartphones or wristbands (Swan, 2012). Second, increasing progress in data analytics and artificial intelligence will help to accurately interpret this sensor data (Müller and Bostrom, 2016). Finally, an increasing degree of connectivity of different devices in our environment will help to share, use and act upon the transmitted information (Mazhelis et al., 2012).

Assuming an accurate and validated detection of passive fatigue in non-sleep-deprived drivers under real-road driving conditions with unobtrusive sensors in future cars, the study results of this thesis provide several specific recommendations for applying thermal stimulation as comfort functions in transportation systems.

Starting with the evidence that the effect of **thermal stimulation is short-lived**, these findings shape the boundaries upon which manufacturers can make use of the effect. The short span of the effect of thermal stimulation suggests that the stimulus has to be repeated in order to reduce fatigue, if the monotony of the drive stays constant. Applying thermal stimulation often, though, will inevitably lead to thermal discomfort.

Furthermore, **leg cooling** or **cooling durations of less than 2 minutes are not recommended** for temperatures of 15 °C or over. This is because no or small physiological effects were seen for these settings of the AC (Chapter 5.2 and 5.3).

Next, it is recommended to **vary the stimulus after the repetition of several thermal stimuli**, as these are subject to habituation and therefore lose their effect on physiological fatigue. The last study of this thesis showed that the stimulus reduced physiological fatigue significantly for the entire cooling period when drivers experienced the stimulus for the first time. The second application of the stimulus in contrast, reduced fatigue for about 2.5 minutes, with minimal effectiveness upon subsequent applications. (Chapter 7). Possible variations can be the use of mental and haptic stimulation as other driving studies have shown that these are effective against fatigue (see Chapter 2.1.2).

Finally, systems must be designed such that **the fatigue countermeasure is offered to the driver** including the reason of detected fatigue, rather than fully automated actions, at least for novice users of the system.

9.2 Directions for Future Research

Several recommendations for the direction of future research activities can be made. The first one addresses the issue of habituation during the drive. As the last study showed, repetitions of the 4-minute stimulus of 15 °C were subject to habituation. Future research should be directed towards an **investigation of the subsequent application of different kinds of sensory stimulation** as a means to prolong the activating effect. As the review in Chapter 2.1.2 showed, a variety of stimulation may reduce passive fatigue and restore vigilance. These stimuli can be of mental, haptic, thermal or olfactory nature. While studies exist that analyzed the effectiveness of one stimulus, there is a deficit of studies investigating different types of consecutive stimuli.

To expand the knowledge on different types of stimulation during passive fatigue, it is also worthwhile to test a **combination of different stimuli**, i.e., multi-sensory stimulation and its possible interaction effects. In learning about these interactions, new opportunities may arise to design a fatigue management system using both single and multi-sensory stimulation, possibly providing enough variety to mitigate the habituation effect occurring upon repetition of a singular stimulus.

The second direction for future research are **long-term studies** in which drivers use a fatigue management system not only for a single drive, as was the case in the studies of this thesis, but also over a longer period of time, e.g., several months. This could deliver important insights, as it is unknown whether habituation to these systems only occurs when the experienced stimulations follow each other within a short amount of time during one drive, or whether habituation also happens when more time lies between drives.

The third important direction for future research which has already been addressed as a limitation of this thesis in Chapter 8.2, are **real road studies on the effect of sensory stimulation**. All driving studies mentioned in Chapter 2.1.2 that applied a mental (Gershon et al., 2009a; Jarosch et al., 2017; Schömig et al., 2015), haptic (Merat and Jamson, 2013; Wang et al., 2017), thermal (Van Veen, 2016) and olfactory (Mahachandra and Garnaby, 2015) stimulus have been conducted in driving simulators. Investigating the effects in real driving conditions would be particularly interesting because the effect of stimulation may be increased since the drivers are less fatigued based on the visuomotor stimulation and the countermeasures might relieve the strain from drivers more efficiently. Conversely, the effect of stimulation could also be decreased, as it may be less noticeable in combination with the visuomotor stimulation. Therefore, on-road driving studies would pose a great contribution to the field of driver fatigue research.

A fourth research direction has already been addressed earlier in this thesis (Chapter XXV and 9.2) and deals with the limited ability to **diagnose the cause of fatigue** in current systems. While the analysis of fatigue symptoms such as changes in steering behavior or gaze and blink behavior provide indications for the presence of fatigue, this analysis is not suited to select appropriate countermeasures against fatigue. This is because the causal factors of fatigue, such as sleep deprivation, task overload or task underload (Figure 2) are crucial for the selection of suited countermeasures (Van Veen et al., 2014). Therefore, not only does the scientific validation of the detection accuracies of fatigue detection systems need improvement, but also their contextual awareness in terms of the causal factors of fatigue.

The discussion of causal factors leads to the last recommendation for future research. As the passive fatigue model (Figure 3) in Chapter 2.1.2 suggests, the causes for monotony and hence passive fatigue stem from three different types of factors: environmental, vehicular and personal factors. Finding solutions within these fields would **counteract the onset of fatigue** before it would be able cause discomfort or impair the driver. Whereas it seems unlikely to hold back the increase of automation in the vehicular field, further research should be spent on reducing the monotony from the environment. Thiffault and Bergeron (2003) investigated whether a road side with random trees, farms and houses was more stimulating than a road side with regularly planted pine trees or a road side covered with only grass. Future research could, for example, **investigate the effect of different visual elements**.

Next to the environmental factors, the **field of personal factors**, such as familiarity, should also be further researched. Charlton and Starkey (2011, 2013) have suggested contextual cues, for example speed cues, and motivational conditions in locations that are likely to be scanned by drivers as countermeasures to be implemented by road safety practitioners. Also in this field, future research should be spent on addressing drivers' familiarity to roads closest to their homes.

Appendix

© Springer Fachmedien Wiesbaden GmbH, part of Springer Nature 2020
E. Schmidt, *Effects of Thermal Stimulation during Passive Driver Fatigue*,
Gestaltung hybrider Mensch-Maschine-Systeme/Designing Hybrid Societies,
https://doi.org/10.1007/978-3-658-28158-8

A1 Questionnaire of Fatigue Induction Study

Original German version:

Stanford Sleepiness Scale

Bitte bewerten Sie Ihre **derzeitige Müdigkeit** indem Sie die entsprechende Zahl ankreuzen.

○ 1   Fühle mich aktiv, vital, voll da, hellwach

○ 2   Habe einen klaren Kopf, bin aber nicht in Top-Form; kann mich konzentrieren

○ 3   Wach, aber entspannt; reagiere, bin aber nicht so ganz da

○ 4   Etwas benommen, schlaff

○ 5   Benommen, verliere das Interesse am Wachbleiben, tranig

○ 6   Schläfrig, benommen, kämpfe mit dem Schlaf; würde mich gerne hinlegen

○ 7   Kämpfe nicht mehr gegen den Schlaf, schlafe gleich ein; traumartige Gedanken

A2 Questionnaires of Cooling Studies

A2.1 Cooling Study Employing 6-Minute Upper Body Cooling

Original German versions:

Karolinska Sleepiness Scale

Bitte bewerten Sie Ihre **derzeitige Müdigkeit** indem Sie die entsprechende Zahl ankreuzen. Benutzen Sie auch die Zwischenstufen.

○ 1   = sehr wach

○ 2

○ 3   = wach

○ 4

○ 5   = weder wach noch müde

○ 6

○ 7   = müde, aber keine Probleme wach zu bleiben

○ 8

○ 9   = sehr müde, große Probleme wach zu bleiben

Activation Deactivation Adjective Checklist

Wie haben Sie sich **während den letzten Minuten der Fahrt** gefühlt?
Kreisen Sie dazu die zutreffende Antwortmöglichkeit ein.

	Fühle ich definitiv	Fühle ich etwas	Unent- schlossen	Fühle ich de- finitiv nicht
1. wachsam	✓✓	✓	?	nein
2. angespannt	✓✓	✓	?	nein
3. elanvoll	✓✓	✓	?	nein
4. friedlich	✓✓	✓	?	nein
5. verkrampft	✓✓	✓	?	nein
6. hellwach	✓✓	✓	?	nein
7. ruhig	✓✓	✓	?	nein
8. lebendig	✓✓	✓	?	nein
9. ängstlich	✓✓	✓	?	nein
10. schlaftrunken	✓✓	✓	?	nein
11. in Ruhe	✓✓	✓	?	nein
12. lebhaft	✓✓	✓	?	nein
13. müde	✓✓	✓	?	nein
14. beruhigt	✓✓	✓	?	nein
15. angestrengt	✓✓	✓	?	nein
16. energetisch	✓✓	✓	?	nein
17. nervös	✓✓	✓	?	nein
18. schläfrig	✓✓	✓	?	nein
19. gelassen	✓✓	✓	?	nein
20. aktiv	✓✓	✓	?	nein

Wärmeempfinden

Descripitive wording for COOL condition:

"Bitte bewerten Sie Ihr eigenes Wärmeempfinden **während der Kühlung** auf der folgenden Skala."

Descripitive wording for CONT condition:

"Bitte bewerten Sie Ihr eigenes Wärmeempfinden **während der Fahrt** auf der folgenden Skala."

Ich fand die Umgebung ...

kalt	kühl	leicht kühl	neutral	leicht warm	warm	heiß
-3	-2	-1	0	+1	+2	+3
○	○	○	○	○	○	○

Thermischer Komfort

Descripitive wording for COOL condition:

"Bitte bewerten Sie Ihr Komfortempfinden **während der Kühlung** auf der folgenden Skala."

Descripitive wording for CONT condition:

"Bitte bewerten Sie Ihr Komfortempfinden **während der Fahrt** auf der folgenden Skala."

Mir persönlich war es ...

viel zu kühl	zu kühl	angenehm kühl	angenehm	angenehm warm	zu warm	viel zu warm
-3	-2	-1	0	+1	+2	+3
O	O	O	O	O	O	O

Eindrücke zur Kühlung

(asked after the COOL-drive)

War die Kühlung angenehm?

O ja   O nein

Begründung

Denken Sie, dass die kurzzeitige Kühlung die Müdigkeit reduziert hat?

○ ja   ○ nein

Begründung

Sonstige Kommentare

(asked after both drives have been completed)

Wie würden Sie eine monotone Fahrt lieber mit oder ohne kurzzeitiger Kühlung austragen?

○ mit Kühlung   ○ ohne Kühlung

Begründung

A2.2 Cooling Study Employing 2 and 4-Minute Upper Body Cooling

Original German versions:

Karolinska Sleepiness Scale

Bitte bewerten Sie Ihre **derzeitige Müdigkeit** indem Sie die entsprechende Zahl ankreuzen. Benutzen Sie auch die Zwischenstufen.

○ 1   = sehr wach

○ 2

○ 3   = wach

○ 4

○ 5   = weder wach noch müde

○ 6

○ 7   = müde, aber keine Probleme wach zu bleiben

○ 8

○ 9   = sehr müde, große Probleme wach zu bleiben

Activation Deactivation Adjective Checklist

Wie haben Sie sich **während den letzten Minuten der Fahrt** gefühlt?
Kreisen Sie dazu die zutreffende Antwortmöglichkeit ein.

	Fühle ich definitiv	Fühle ich etwas	Unent- schlossen	Fühle ich de- finitiv nicht
1. wachsam	✓✓	✓	?	nein
2. angespannt	✓✓	✓	?	nein
3. elanvoll	✓✓	✓	?	nein
4. friedlich	✓✓	✓	?	nein
5. verkrampft	✓✓	✓	?	nein
6. hellwach	✓✓	✓	?	nein
7. ruhig	✓✓	✓	?	nein
8. lebendig	✓✓	✓	?	nein
9. ängstlich	✓✓	✓	?	nein
10. schlaftrunken	✓✓	✓	?	nein
11. in Ruhe	✓✓	✓	?	nein
12. lebhaft	✓✓	✓	?	nein
13. müde	✓✓	✓	?	nein
14. beruhigt	✓✓	✓	?	nein
15. angestrengt	✓✓	✓	?	nein
16. energetisch	✓✓	✓	?	nein
17. nervös	✓✓	✓	?	nein
18. schläfrig	✓✓	✓	?	nein
19. gelassen	✓✓	✓	?	nein
20. aktiv	✓✓	✓	?	nein

Wärmeempfinden

Descripitive wording for COOL condition:

"Bitte bewerten Sie Ihr eigenes Wärmeempfinden **während der Kühlung** auf der folgenden Skala."

Descripitive wording for CONT condition:

"Bitte bewerten Sie Ihr eigenes Wärmeempfinden **während der Fahrt** auf der folgenden Skala."

Ich fand die Umgebung ...

kalt	kühl	leicht kühl	neutral	leicht warm	warm	heiß
-3	-2	-1	0	+1	+2	+3
O	O	O	O	O	O	O

Thermischer Komfort

Descripitive wording for COOL condition:

"Bitte bewerten Sie Ihr Komfortempfinden **während der Kühlung** auf der folgenden Skala."

Descripitive wording for CONT condition:

"Bitte bewerten Sie Ihr Komfortempfinden **während der Fahrt** auf der folgenden Skala."

Mir persönlich war es …

viel zu kühl	zu kühl	angenehm kühl	angenehm	angenehm warm	zu warm	viel zu warm
-3	-2	-1	0	+1	+2	+3
○	○	○	○	○	○	○

Eindrücke zur Kühlung

(asked after the COOL-drive)

War die Kühlung angenehm?

○ ja   ○ nein

Begründung

Denken Sie, dass die kurzzeitige Kühlung die Müdigkeit reduziert hat?

○ ja   ○ nein

Begründung

Sonstige Kommentare

(asked after both drives have been completed)

Wie würden Sie eine monotone Fahrt lieber mit oder ohne kurzzeitiger Kühlung austragen?

○ mit Kühlung   ○ ohne Kühlung

Begründung

A2.3 Cooling Study Employing 4-Minute Lower Leg Cooling

Original German versions:

Karolinska Sleepiness Scale

Bitte bewerten Sie Ihre **derzeitige Müdigkeit** indem Sie die entsprechende Zahl ankreuzen. Benutzen Sie auch die Zwischenstufen.

○	1	= sehr wach
○	2	
○	3	= wach
○	4	
○	5	= weder wach noch müde
○	6	
○	7	= müde, aber keine Probleme wach zu bleiben
○	8	
○	9	= sehr müde, große Probleme wach zu bleiben

Activation Deactivation Adjective Checklist

Wie haben Sie sich **während den letzten Minuten der Fahrt** gefühlt?
Kreisen Sie dazu die zutreffende Antwortmöglichkeit ein.

	Fühle ich definitiv	Fühle ich etwas	Unent-schlossen	Fühle ich definitiv nicht
1. wachsam	✓✓	✓	?	nein
2. angespannt	✓✓	✓	?	nein
3. elanvoll	✓✓	✓	?	nein
4. friedlich	✓✓	✓	?	nein
5. verkrampft	✓✓	✓	?	nein
6. hellwach	✓✓	✓	?	nein
7. ruhig	✓✓	✓	?	nein
8. lebendig	✓✓	✓	?	nein
9. ängstlich	✓✓	✓	?	nein
10. schlaftrunken	✓✓	✓	?	nein
11. in Ruhe	✓✓	✓	?	nein
12. lebhaft	✓✓	✓	?	nein
13. müde	✓✓	✓	?	nein
14. beruhigt	✓✓	✓	?	nein
15. angestrengt	✓✓	✓	?	nein
16. energetisch	✓✓	✓	?	nein
17. nervös	✓✓	✓	?	nein
18. schläfrig	✓✓	✓	?	nein
19. gelassen	✓✓	✓	?	nein
20. aktiv	✓✓	✓	?	nein

Thermischer Komfort

Descripitive wording for COOL group:

"Bitte bewerten Sie Ihr Komfortempfinden **während der Kühlung** auf der folgenden Skala."

Descripitive wording for CONT group:

"Bitte bewerten Sie Ihr Komfortempfinden **während der Fahrt** auf der folgenden Skala."

Ich fand die Umgebung …

kalt	kühl	leicht kühl	neutral	leicht warm	warm	heiß
-3	-2	-1	0	+1	+2	+3
O	O	O	O	O	O	O

Eindrücke zur Kühlung

(asked in the COOL group only)

War die Kühlung angenehm?

O ja   O nein

Begründung

Denken Sie, dass die kurzzeitige Kühlung die Müdigkeit reduziert hat?

○ ja   ○ nein

Begründung

A3 Questionnaires of Closed-Loop Study

Original German versions:

Karolinska Sleepiness Scale

Bitte bewerten Sie Ihre **derzeitige Müdigkeit** indem Sie die entsprechende Zahl ankreuzen. Benutzen Sie auch die Zwischenstufen.

○ 1   = sehr wach

○ 2

○ 3   = wach

○ 4

○ 5   = weder wach noch müde

○ 6

○ 7   = müde, aber keine Probleme wach zu bleiben

○ 8

○ 9   = sehr müde, große Probleme wach zu bleiben

Activation Deactivation Adjective Checklist

Wie haben Sie sich **während den letzten Minuten der Fahrt** gefühlt? Kreisen Sie dazu die zutreffende Antwortmöglichkeit ein.

		Fühle ich definitiv	Fühle ich etwas	Unent- schlossen	Fühle ich de- finitiv nicht
1.	wachsam	✓✓	✓	?	nein
2.	angespannt	✓✓	✓	?	nein
3.	elanvoll	✓✓	✓	?	nein
4.	friedlich	✓✓	✓	?	nein
5.	verkrampft	✓✓	✓	?	nein
6.	hellwach	✓✓	✓	?	nein
7.	ruhig	✓✓	✓	?	nein
8.	lebendig	✓✓	✓	?	nein
9.	ängstlich	✓✓	✓	?	nein
10.	schlaftrunken	✓✓	✓	?	nein
11.	in Ruhe	✓✓	✓	?	nein
12.	lebhaft	✓✓	✓	?	nein
13.	müde	✓✓	✓	?	nein
14.	beruhigt	✓✓	✓	?	nein
15.	angestrengt	✓✓	✓	?	nein
16.	energetisch	✓✓	✓	?	nein
17.	nervös	✓✓	✓	?	nein
18.	schläfrig	✓✓	✓	?	nein
19.	gelassen	✓✓	✓	?	nein
20.	aktiv	✓✓	✓	?	nein

Van der Laan Acceptance Scale

Descriptive wording for baseline and notification group:

"Bitte beurteilen Sie **die Möglichkeit automatisiert erfrischt zu werden auf Basis der heute erlebten Fahrt** mit Hilfe der folgenden Wortpaare. Setzen Sie dazu jeweils ein Kreuz pro Zeile, um zu verdeutlichen, **zu welchem der zwei Wörter Sie tendieren.**

Automatisierte Erfrischung finde ich…"

Descriptive wording for recommendation group:

"Bitte beurteilen Sie die **Möglichkeit des automatisierten Vorschlages, erfrischt zu werden, auf Basis der heute erlebten Fahrt** mit Hilfe der folgenden Wortpaare. Setzen Sie dazu jeweils ein Kreuz pro Zeile, um zu verdeutlichen, **zu welchem der zwei Wörter Sie tendieren.**

Den automatisierten Vorschlag zur Erfrischung finde ich…"

1.	nützlich	O	O	O	O	O	nutzlos
2.	angenehm	O	O	O	O	O	unangenehm
3.	schlecht	O	O	O	O	O	gut
4.	nett	O	O	O	O	O	nervig
5.	wirksam	O	O	O	O	O	unnötig
6.	ärgerlich	O	O	O	O	O	erfreulich
7.	hilfreich	O	O	O	O	O	wertlos
8.	nicht wünsch-enswert	O	O	O	O	O	wünschenswert
9.	aktivierend	O	O	O	O	O	einschläfernd

Automation Trust

Descriptive wording for baseline and notification group:

"Bitte bewerten Sie Ihr Vertrauen gegenüber dem **automatisierten Erfrischen als technisches System.**"

Descriptive wording for recommendation group:

"Bitte bewerten Sie Ihr Vertrauen gegenüber dem **automatisierten Vorschlag zur Erfrischung als technisches System.**"

Bitte geben Sie den **Grad Ihrer Zustimmung** zu jeder Aussage an. **Markieren Sie dazu in jeder Zeile eine Position** zwischen „stimme überhaupt nicht zu" und „stimme vollkommen zu".

Stimme über-
haupt nicht zu

Stimme voll-
kommen zu

1. Ich kann dem System vertrauen. ① ② ③ ④ ⑤ ⑥ ⑦

2. Das System ist irreführend. ① ② ③ ④ ⑤ ⑥ ⑦

3. Die Aktionen des Systems sind undurchsichtig. ① ② ③ ④ ⑤ ⑥ ⑦

4. Ich misstraue den Aktionen, Absichten oder Konsequenzen des Systems. ① ② ③ ④ ⑤ ⑥ ⑦

5. Ich bin dem System gegenüber wachsam. ① ② ③ ④ ⑤ ⑥ ⑦

6. Ich bin mit dem System vertraut. ① ② ③ ④ ⑤ ⑥ ⑦

7. Die Aktionen des Systems führen zu nachteiligen oder schädlichen Konsequenzen. ① ② ③ ④ ⑤ ⑥ ⑦

8. Ich traue mir zu, das System zu nutzen. ① ② ③ ④ ⑤ ⑥ ⑦

9. Das System ist glaubwürdig. ① ② ③ ④ ⑤ ⑥ ⑦

10. Ich kann mich auf das System verlassen. ① ② ③ ④ ⑤ ⑥ ⑦

11. Das System bietet Sicherheit. ① ② ③ ④ ⑤ ⑥ ⑦

12. Das System ist zuverlässig. ① ② ③ ④ ⑤ ⑥ ⑦

User Experience Questionnaire

Bitte geben Sie Ihre Beurteilung ab.

Um die Anzeige zu bewerten, füllen Sie bitte den nachfolgenden Fragebogen aus. Er besteht aus Gegensatzpaaren von Eigenschaften, die die Anzeige haben kann. Abstufungen zwischen den Gegensätzen sind durch Kreise dargestellt. Durch Ankreuzen eines dieser Kreise können Sie Ihre Zustimmung zu einem Begriff äußern.

Beispiel:

| attraktiv | $\bigcirc$ | $\otimes$ | $\bigcirc$ | $\bigcirc$ | $\bigcirc$ | $\bigcirc$ | $\bigcirc$ | unattraktiv |

Mit dieser Beurteilung sagen Sie aus, dass Sie die Anzeige eher attraktiv als unattraktiv einschätzen.

Entscheiden Sie möglichst spontan. Es ist wichtig, dass Sie nicht lange über die Begriffe nachdenken, damit Ihre unmittelbare Einschätzung zum Tragen kommt.

Bitte kreuzen Sie immer eine Antwort an, auch wenn Sie bei der Einschätzung zu einem Begriffspaar unsicher sind oder finden, dass es nicht so gut zur Anzeige passt.

Es gibt keine „richtige" oder „falsche" Antwort. Ihre persönliche Meinung zählt!

Bitte geben Sie nun Ihre Einschätzung zur Anzeige ab. Kreuzen Sie bitte nur einen Kreis pro Zeile an.

	1	2	3	4	5	6	7		
unverständlich	O	O	O	O	O	O	O	verständlich	1
kreativ	O	O	O	O	.O	O	O	phantasielos	2
leicht zu lernen	O	O	O	O	O	O	O	schwer zu lernen	3
wertvoll	O	O	O	O	O	O	O	minderwertig	4
langweilig	O	O	O	O	O	O	O	spannend	5
uninteressant	O	O	O	O	O	O	O	interessant	6
schnell	O	O	O	O	O	O	O	langsam	7
originell	O	O	O	O	O	O	O	konventionell	8
gut	O	O	O	O	O	O	O	schlecht	9
kompliziert	O	O	O	O	O	O	O	einfach	10
herkömmlich	O	O	O	O	O	O	O	neuartig	11
unangenehm	O	O	O	O	O	O	O	angenehm	12
aktivierend	O	O	O	O	O	O	O	einschläfernd	13
ineffizient	O	O	O	O	O	O	O	effizient	14
übersichtlich	O	O	O	O	O	O	O	verwirrend	15
unpragmatisch	O	O	O	O	O	O	O	pragmatisch	16
aufgeräumt	O	O	O	O	O	O	O	überladen	17
attraktiv	O	O	O	O	O	O	O	unattraktiv	18
unsympathisch	O	O	O	O	O	O	O	sympathisch	19
konservativ	O	O	O	O	O	O	O	innovativ	20

Eindrücke zur Kühlung

War die Kühlung angenehm?

○ ja   ○ nein

Begründung

Denken Sie, dass die kurzzeitige Kühlung die Müdigkeit reduziert hat?

○ ja   ○ nein

Begründung

A4 Simulink Model for Fatigue Detection

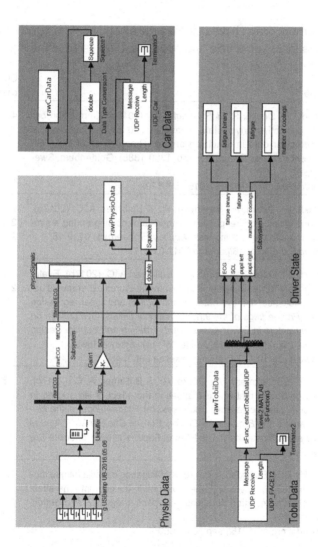

Figure A-1: Simulink model for fatigue detection
Source: *Own illustration*

A5 Author's Work

The above monograph bases on the articles and posters listed below. Thesis chapters including article and poster text are listed in the right column.

	Article	Chapters including text
I	Schmidt, E., Decke, R., & Rasshofer, R. (2016a). Correlation between subjective driver state measures and psychophysiological and vehicular data in simulated driving. *Proceedings of the IEEE Intelligent Vehicles Symposium* (pp. 1380-1385). Gothenburg, Sweden: IEEE. https://doi.org/10.1109/IVS.2016.7535570.	4
II	Schmidt, E., Decke, R., Rasshofer, R., & Bullinger, A. C. (2017a). Psychophysiological responses to short-term cooling during a simulated monotonous driving task. *Applied Ergonomics, 62*, 9-18. https://doi.org/10.1016/j.apergo.2017.01.017.	2.1.1, 2.1.3, 2.2.1, 2.2.2, 2.2.4, 3, 5.1
III	Schmidt, E., Ochs, J., Decke, R., & Bullinger, A. C. (2017b). Evaluating drivers' states in sleepiness countermeasures experiments using physiological and eye data - hybrid logistic and linear regression model. *Proceedings of the 9th International Driving Symposium on Human Factors in Driver Assessment, Training and Vehicle Design* (pp. 284-290). Manchester Village, VT: University of Iowa. https://doi.org/10.17077/driving assessment.1648.	2.1, 2.1.1, 2.1.4, 6
IV	Schmidt, E., Dettmann, A., Decke, R., & Bullinger, A. C. (2017c). Cold legs do not matter – investigating the effect of leg cooling to overcome passive fatigue. *Postersession Proceedings of the Human Factors and Ergonomics Society Europe Chapter 2017 Annual Conference*. Rome, Italy: The Human Factors and Ergonomics Society, Europe Chapter.	5.3
V	Schmidt, E., & Bullinger, A. C. (2019). Mitigating passive fatigue during monotonous drives with thermal stimuli: Insights into the effect of different stimulation durations. *Accident Analysis & Prevention, 126*, 115-121. https://doi.org/10.1016/j.aap.2017.12.005.	1.1, 2.1, 2.1.1, 2.1.2, 2.2.1, 2.2.2, 2.2.3, 2.2.4, 5.2

References

Acosta, M. C., Gallar, J., & Belmonte, C. (1999). The influence of eye solutions on blinking and ocular comfort at rest and during work at video
display terminals. *Experimental Eye Research, 68*(6), 663-669.
https://doi.org/10.1006/exer.1998.0656.

Åkerstedt, T., & Gillberg, M. (1990). Subjective and objective sleepiness in
the active individual. *International Journal of Neuroscience, 52*(1-2),
29-37.
https://doi.org/10.3109/00207459008994241.

Anund, A., Kecklund, G., Peters, B., & Åkerstedt, T. (2008). Driver sleepiness and individual differences in preferences for countermeasures.
Journal of Sleep Research, 17(1), 16-22.
https://doi.org/10.1111/j.1365-2869.2008.00633.x.

ASHRAE (1966). *Thermal Comfort Conditions*. New York: ASHRAE Standard 55-1966.

Audi AG (2018). *The car with empathy*. Retrieved from https://audi-en-
counter.com/en/audi-fit-driver, Access Date: 07.08.2018.

Backs, R. W., da Silva, S. P., & Han, K. (2005). A comparison of younger
and older adults' self-assessment manikin ratings of affective pictures.
Experimental Aging Research, 31(4), 421-440.
https://doi.org/10.1080/03610730500206808.

Balkin, T. J., Horrey, W. J., Graeber, R. C., Czeisler, C. A., & Dinges, D. F.
(2011). The challenges and opportunities of technological approaches
to fatigue management. *Accident Analysis & Prevention, 43*(2), 565-
572.
https://doi.org/10.1016/j.aap.2009.12.006.

© Springer Fachmedien Wiesbaden GmbH, part of Springer Nature 2020
E. Schmidt, *Effects of Thermal Stimulation during Passive Driver Fatigue*,
Gestaltung hybrider Mensch-Maschine-Systeme/Designing Hybrid Societies,
https://doi.org/10.1007/978-3-658-28158-8

Beatty, J. (1982). Task-evoked pupillary responses, processing load, and the structure of processing resources. *Psychological Bulletin, 91*(2), 276-292.

Bedford, T. (1936). *The warmth factor in comfort at work. A Physiological Study of Heating and Ventilation.* Med. Res. Council, Industr. Health Res. Board, Report No 76. London: HMSO.

Beggiato, M. (2015). *Changes in motivational and higher level cognitive processes when interacting with in-vehicle automation.* PhD thesis. Technische Universität Chemnitz.

BMW AG (2019). Caring Car. Retrieved from https://www.bmw.de/de/topics/faszination-bmw/connecteddrive/bmw-connected-drive-uebersicht.html, Access Date: 10.06.2019.

Bradley, M. M., & Lang, P. J. (1994). Measuring emotion: the self-assessment manikin and the semantic differential. *Journal of Behavior Therapy and Experimental Psychiatry, 25*(1), 49-59. https://doi.org/10.1016/0005-7916(94)90063-9.

Bradley, M. M., Miccoli, L., Escrig, M. A., & Lang, P. J. (2008). The pupil as a measure of emotional arousal and autonomic activation. *Psychophysiology, 45*(4), 602-607. https://doi.org/10.1111/j.1469-8986.2008.00654.x.

Bradley, M. M. (2009). Natural selective attention: Orienting and emotion. *Psychophysiology, 46*(1), 1-11. https://doi.org/10.1111/j.1469-8986.2008.00702.x.

Briest, S., Karrer, K., & Schleicher, R. (2006). Driving without awareness: Examination of the phenomenon. *Vision in Vehicles XI*, 89-141.

Brown, I. D., (1994). Driver fatigue. *Human Factors 36*(2), 298-314. https://doi.org/10.1177/001872089403600210.

Brüggemann, W. (1980*). Kneipptherapie: Ein Lehrbuch.* Berlin: Springer. https://doi.org/10.1007/978-3-642-96577-7.

Bubb, H., Bengler, K., Grünen, R. E., & Vollrath, M. (Eds.) (2015). *Automobilergonomie*. Wiesbaden: Springer Vieweg.

Bullinger-Hoffmann, A. C., & Mühlstedt, J. (Eds.). (2016). *Homo Sapiens Digitalis-Virtuelle Ergonomie und digitale Menschmodelle*. Wiesbaden: Springer Vieweg.

Bundele, M. M., & Banerjee, R. (2009). Detection of fatigue of vehicular driver using skin conductance and oximetry pulse: a neural network approach. *Proceedings of the 11th International Conference on Information Integration and Web-based Applications & Services* (pp. 739-744). New York, NY: ACM.
https://doi.org/10.1145/1806338.1806478.

Casagrande, M., Violani, C., Curcio, G., & Bertini, M. (1997). Assessing vigilance through a brief pencil and paper letter cancellation task (LCT): effects of one night of sleep deprivation and of the time of day. *Ergonomics, 40*(6), 613-630.
https://doi.org/10.1080/001401397187919.

Charlton, S. G., & Starkey, N. J. (2011). Driving without awareness: The effects of practice and automaticity on attention and driving. *Transportation Research Part F: Traffic Psychology and Behaviour, 14*(6), 456-471.
https://doi.org/10.1016/j.trf.2011.04.010.

Charlton, S. G., & Starkey, N. J. (2013). Driving on familiar roads: Automaticity and inattention blindness. *Transportation Research Part F: Traffic Psychology and Behaviour, 19*, 121-133.
https://doi.org/10.1016/j.trf.2013.03.008.

Cohen R.A. (2011). Yerkes–Dodson Law. In: J.S. Kreutzer, J. DeLuca, & B. Caplan (Eds.), *Encyclopedia of Clinical Neuropsychology*. New York, NY: Springer.
https://doi.org/10.1007/978-0-387-79948-3_1340.

Collins, K. J., Abdel-Rahman, T. A., Easton, J. C., Sacco, P., Ison, J., & Dore, C. J. (1996). Effects of facial cooling on elderly and young subjects: interactions with breath-holding and lower body negative pressure. *Clinical Science, 90*(6), 485-492. https://doi.org/10.1042/cs0900485.

Connor, J., Whitlock, G., Norton, R., & Jackson, R. (2001). The role of driver sleepiness in car crashes: a systematic review of epidemiological studies. *Accident Analysis & Prevention, 33*(1), 31-41. https://doi.org/10.1016/S0001-4575(00)00013-0.

Coughlin, J. F., Reimer, B., & Mehler, B. (2009). Driver wellness, safety & the development of an awarecar. AgeLab, MIT.

Cramer, H., Evers, V., Ramlal, S., Van Someren, M., Rutledge, L., Stash, N., Aroyo, L., & Wielinga, B. (2008). The effects of transparency on trust in and acceptance of a content-based art recommender. *User Modeling and User-Adapted Interaction, 18*(5), 455-496. https://doi.org/10.1007/s11257-008-9051-3.

Curcio, G., Casagrande, M., & Bertini, M. (2001). Sleepiness: evaluating and quantifying methods. *International Journal of Psychophysiology, 41*(3), 251-263. https://doi.org/10.1016/S0167-8760(01)00138-6.

Daanen, H. A., van de Vliert, E., & Huang, X. (2003). Driving performance in cold, warm, and thermoneutral environments. *Applied Ergonomics, 34*(6), 597-602. https://doi.org/10.1016/S0003-6870(03)00055-3.

Daimler AG (2018). *ENERGIZING Komfortsteuerung: Wellness beim Fahren*. Retrieved from http://media.daimler.com/marsMediaSite/de/instance/ko/ENERGIZING-Komfortsteuerung-Wellness-beim-Fahren.xhtml?oid=22934464, Access Date: 07.08.2018.

D'Angelo, L. T., & Lüth, T. C. (2011). Integrated systems for distraction-free vital signs measurement in vehicles. *ATZ worldwide eMagazine, 113*(11), 52-56. https://doi.org/10.1365/s38311-011-0116-2.

Dawson, D., Searle, A. K., & Paterson, J. L. (2014). Look before you (s)leep: evaluating the use of fatigue detection technologies within a fatigue risk management system for the road transport industry. *Sleep Medicine Reviews, 18*(2), 141-152. https://doi.org/10.1016/j.smrv.2013.03.003.

DePaoli, L. C., & Sweeney, D. C. (2000). Further validation of the positive and negative affect schedule. *Journal of Social Behavior and Personality, 15*(4), 561-568.

Desmond, P. A., & Matthews, G. (1997). Implications of task-induced fatigue effects for in-vehicle countermeasures to driver fatigue. *Accident Analysis & Prevention, 29*(4), 515-523. https://doi.org/10.1016/S0001-4575(97)00031-6.

De Wijk, R. A., & Zijlstra, S. M. (2012). Differential effects of exposure to ambient vanilla and citrus aromas on mood, arousal and food choice. *Flavour, 1*(1), 24. https://doi.org/10.1186/2044-7248-1-24.

Dingus, T. A., Klauer, S. G., Neale, V. L., Petersen, A., Lee, S. E., Sudweeks, J. D., et al. (2006). *The 100-car naturalistic driving study, Phase II-results of the 100-car field experiment* (No. DOT HS-810 593). Washington, DC: U.S. Department of Transportation.

Dingus, T. A., Guo, F., Lee, S., Antin, J. F., Perez, M., Buchanan-King, M., & Hankey, J. (2016). Driver crash risk factors and prevalence evaluation using naturalistic driving data. *Proceedings of the National Academy of Sciences, 113*(10), 2636-2641. https://doi.org/10.1073/pnas.1513271113.

Eglund, N. (1982). Spectral analysis of heart rate variability as an indicator of driver fatigue. *Ergonomics, 25*(7), 663-672. https://doi.org/10.1080/00140138208925026.

Eilebrecht, B., Wartzek, T., Lem, J., Vogt, R., & Leonhardt, S. (2011). Capacitive electrocardiogram measurement system in the driver seat. *ATZ worldwide eMagazine, 113*(3), 50-55. https://doi.org/10.1365/s38311-011-0034-3.

Eilers, K., Nachreiner, F., & Hänecke, K. (1986). Entwicklung und Überprüfung einer Skala zur Erfassung subjektiv erlebter Anstrengung. *Zeitschrift für Arbeitswissenschaft, 40*, 214-224.

Ekkekakis, P., Hall, E. E., & Petruzzello, S. J. (2005). Evaluation of the circumplex structure of the Activation Deactivation Adjective Check List before and after a short walk. *Psychology of Sport and Exercise, 6*(1), 83-101. https://doi.org/10.1016/j.psychsport.2003.10.005.

Ekman, P., & Rosenberg, E. L. (1997). What the face reveals: Basic and applied studies of spontaneous expression using the Facial Action Coding System (FACS). New York, NY: Oxford University Press.

El Falou, W., Duchêne, J., Grabisch, M., Hewson, D., Langeron, Y., & Lino, F. (2003). Evaluation of driver discomfort during long-duration car driving. *Applied Ergonomics, 34*(3), 249-255. https://doi.org/10.1016/S0003-6870(03)00011-5.

EN ISO 14505-3 (2006). Ergonomics of the thermal environment - Evaluation of the thermal environment in vehicles - Part 3: Evaluation of thermal comfort using human subjects.

Etikan, I., Musa, S. A., & Alkassim, R. S. (2016). Comparison of convenience sampling and purposive sampling. *American Journal of Theoretical and Applied Statistics, 5*(1), 1-4. https://doi.org/10.11648/j.ajtas.20160501.11.

Farahmand, B., & Boroujerdian, A. M. (2018). Effect of road geometry on driver fatigue in monotonous environments: A simulator study. *Transportation Research Part F: Traffic Psychology and Behaviour, 58*, 640-651. https://doi.org/10.1016/j.trf.2018.06.021.

Fell, D. L., & Black, B. (1997). Driver fatigue in the city. *Accident Analysis & Prevention, 29*(4), 463-469. https://doi.org/10.1016/S0001-4575(97)00025-0.

Fletcher, L., Petersson, L., & Zelinsky, A. (2005). Road scene monotony detection in a fatigue management driver assistance system. *Proceedings of the IEEE Intelligent Vehicles Symposium* (pp. 484-489). Las Vegas, NV: IEEE. https://doi.org/10.1109/IVS.2005.1505150.

Fors, C., Ahlstrom, C., & Anund, A. (2016). A comparison of driver sleepiness in the simulator and on the real road. *Journal of Transportation Safety & Security, 10*(1-2), 72-87. https://doi.org/10.1080/19439962.2016.1228092.

Fram, E. H., & McCarthy, M. S. (2003). From employee to brand champion. *Marketing Management, 12*(1), 24-29.

Friedrichs, F., & Yang, B. (2010). Camera-based drowsiness reference for driver state classification under real driving conditions. *Proceedings of the IEEE Intelligent Vehicles Symposium* (pp. 101-106). San Diego, CA: IEEE. https://doi.org/10.1109/IVS.2010.5548039.

Frith, C. D., & Allen, H. A. (1983). The skin conductance orienting response as an index of attention. *Biological Psychology, 17*(1), 27-39. https://doi.org/10.1016/0301-0511(83)90064-9.

Ganz, C. (2016). Die Kneipp'sche Gesundheitslehre. *Schweizerische Zeitschrift für Ganzheitsmedizin/Swiss Journal of Integrative Medicine, 28*(3), 148-152. https://doi.org/10.1159/000446130.

Gershon, P., Ronen, A., Oron-Gilad, T., & Shinar, D. (2009a). The effects of an interactive cognitive task (ICT) in suppressing fatigue symptoms in driving. *Transportation Research Part F: Traffic Psychology and Behaviour, 12*(1), 21-28. https://doi.org/10.1016/j.trf.2008.06.004.

Gershon, P., Shinar, D., & Ronen, A. (2009b). Evaluation of experience-based fatigue countermeasures. *Accident Analysis & Prevention, 41*(5), 969-975. https://doi.org/10.1016/j.aap.2009.05.012.

Gershon, P., Shinar, D., Oron-Gilad, T., Parmet, Y., & Ronen, A. (2011). Usage and perceived effectiveness of fatigue countermeasures for professional and nonprofessional drivers. *Accident Analysis & Prevention, 43*(3), 797-803.
https://doi.org/10.1016/j.aap.2010.10.027.

Gillberg, M., Kecklund, G., & Åkerstedt, T. (1994). Relations between performance and subjective ratings of sleepiness during a night awake. *Sleep, 17*(3), 236-241.
https://doi.org/10.1093/sleep/17.3.236.

Goldwater, B. C. (1972). Psychological significance of pupillary movements. *Psychological Bulletin, 77*(5), 340-355.
https://doi.org/10.1037/h0032456.

Goverdovsky, V., Looney, D., Kidmose, P., & Mandic, D. P. (2016). In-ear EEG from viscoelastic generic earpieces: robust and unobtrusive 24/7 monitoring. *IEEE Sensors Journal, 16*(1), 271-277.
https://doi.org/10.1109/JSEN.2015.2471183.

Graham, F. K., & Clifton, R. K. (1966). Heart-rate change as a component of the orienting response. *Psychological Bulletin, 65*(5), 305-320.
https://doi.org/10.1037/h0023258.

Greco, A., Lanata, A., Valenza, G., Scilingo, E. P., & Citi, L. (2014). Electrodermal activity processing: A convex optimization approach. *Proceedings of the 36th Annual International Conference of the Engineering in Medicine and Biology Society* (pp. 2290-2293). Chicago, IL: IEEE.
https://doi.org/10.1109/EMBC.2014.6944077.

G.tec Medical Engineering GmbH (2018). *g.Sensors and Utilities – Instructions for Use.* Retrieved from http://www.gtec.at/Products/Electrodes-and-Sensors/g.Sensors-Specs-Features, Access Date: 19.08.2018.

Hall, M., Frank, E., Holmes, G., Pfahringer, B., Reutemann, P., & Witten, I. H. (2009). The WEKA data mining software: an update. *ACM SIGKDD explorations newsletter, 11*(1), 10-18.
https://doi.org/10.1145/1656274.1656278.

Hallvig, D., Anund, A., Fors, C., Kecklund, G., Karlsson, J. G., Wahde, M., & Åkerstedt, T. (2013). Sleepy driving on the real road and in the simulator – A comparison. *Accident Analysis & Prevention, 50*, 44-50. https://doi.org/10.1016/j.aap.2012.09.033.

Hart, S. G., & Staveland, L. E. (1988). Development of NASA-TLX (Task Load Index): Results of empirical and theoretical research. *Advances in Psychology, 52*, 139-183. https://doi.org/10.1016/S0166-4115(08)62386-9.

Hassenzahl, M. (2007). The hedonic/pragmatic model of user experience. *Towards a UX manifesto*, Lancaster, UK: COST294-MAUSE.

Hayward, J. M., Holmes, W. F., & Gooden, B. A. (1976). Cardiovascular responses in man to a stream of cold air. *Cardiovascular Research, 10*(6), 691-696. https://doi.org/10.1093/cvr/10.6.691.

Heath, M. E., & Downey, J. A. (1990). The cold face test (diving reflex) in clinical autonomic assessment: methodological considerations and repeatability of responses. *Clinical Science, 78*(2), 139-147. https://doi.org/10.1042/cs0780139.

Heindl, S., Struck, J., Wellhöner, P., Sayk, F., & Dodt, C. (2004). Effect of facial cooling and cold air inhalation on sympathetic nerve activity in men. *Respiratory Physiology & Neurobiology, 142*(1), 69-80. https://doi.org/10.1016/j.resp.2004.05.004.

Helton, W. S., & Russell, P. N. (2012). Brief mental breaks and content-free cues may not keep you focused. *Experimental Brain Research, 219*(1), 37-46. https://doi.org/10.1007/s00221-012-3065-0.

Hensel, H. (1981). *Thermoreception and Temperature Regulation.* London: Academic Press.

Hergeth, S. (2016). *Automation trust in conditional automated driving systems: Approaches to Operationalization and Design.* PhD thesis. Technische Universität Chemnitz.

Herscovitch, J., & Broughton, R. (1981). Sensitivity of the Stanford sleepiness scale to the effects of cumulative partial sleep deprivation and recovery oversleeping. *Sleep, 4*(1), 83-92.
https://doi.org/10.1093/sleep/4.1.83.

Hertzberg, H. T. E. (1972). The human buttocks in sitting: pressures, patterns, and palliatives (No. 720005). SAE Technical Paper.
https://doi.org/10.4271/720005.

Hill, S. G., Iavecchia, H. P., Byers, J. C., Bittner, A. C., Zaklade, A. L., & Christ, R. E. (1992). Comparison of four subjective workload rating scales. *Human Factors, 34*(4), 429-439.
https://doi.org/10.1177/001872089203400405.

Hoddes, E., Zarcone, V., Smythe, H., Phillips, R., & Dement, W. C. (1973). Quantification of sleepiness: a new approach. *Psychophysiology, 10*(4), 431-436.
https://doi.org/10.1111/j.1469-8986.1973.tb00801.x.

Holbrook, M. B., & Gardner, M. P. (1993). An approach to investigating the emotional determinants of consumption durations: why do people consume what they consume for as long as they consume it?. *Journal of Consumer Psychology, 2*(2), 123-142.
https://doi.org/10.1016/S1057-7408(08)80021-6.

Horne, J. A., & Reyner, L. A. (1995). Sleep related vehicle accidents. *British Medical Journal, 310*(6979), 565-567.
https://doi.org/10.1136/bmj.310.6979.565.

Hu, S., & Zheng, G. (2009). Driver drowsiness detection with eyelid related parameters by Support Vector Machine. *Expert Systems with Applications, 36*(4), 7651-7658.
https://doi.org/10.1016/j.eswa.2008.09.030.

Hygge, S. (1992). Heat and performance. In D.M. Jones, & A.P. Smith (Eds.), *Handbook of Human Performance, vol. 1. The Physical Environment* (pp. 79-104). London: Academic Press.

Igasaki, T., Nagasawa, K., Murayama, N., & Hu, Z. (2015). Drowsiness estimation under driving environment by heart rate variability and/or breathing rate variability with logistic regression analysis. *Proceedings of the 8th International Conference on Biomedical Engineering and Informatics* (pp. 189-193). Shenyang, China: IEEE. https://doi.org/10.1109/BMEI.2015.7401498.

Imhof, M. (1998). Erprobung der deutschen Version der Adjektiv-Checkliste nach Thayer (1989) zur Erfassung der aktuellen Aktiviertheit. *Zeitschrift für Differentielle und Diagnostische Psychologie, 19*(3), 179-186.

Jagannath, M., & Balasubramanian, V. (2014). Assessment of early onset of driver fatigue using multimodal fatigue measures in a static simulator. *Applied Ergonomics, 45*(4), 1140-1147. https://doi.org/10.1016/j.apergo.2014.02.001.

Janský, L., Vavra, V., Janský, P., Kunc, P., Knížková, I., Jandova, D., & Slováček, K. (2003). Skin temperature changes in humans induced by local peripheral cooling. Journal of *Thermal Biology, 28*(5), 429-437. https://doi.org/10.1016/S0306-4565(03)00028-7.

Jarosch, O., Kuhnt, M., Paradies, S., & Bengler, K. (2017). It's Out of Our Hands Now! Effects of Non-Driving Related Tasks during Highly Automated Driving on Drivers' Fatigue. *Proceedings of the 9th International Driving Symposium on Human Factors in Driver Assessment, Training and Vehicle Design* (pp. 319-325). Manchester Village, VT: University of Iowa. https://doi.org/10.17077/drivingassessment.1653.

Ji, Q., Zhu, Z., & Lan, P. (2004). Real-time nonintrusive monitoring and prediction of driver fatigue. *IEEE Transactions on Vehicular Technology, 53*(4), 1052-1068. https://doi.org/10.1109/TVT.2004.830974.

Jian, J. Y., Bisantz, A. M., & Drury, C. G. (2000). Foundations for an empirically determined scale of trust in automated systems. *International Journal of Cognitive Ergonomics, 4*(1), 53-71. https://doi.org/10.1207/S15327566IJCE0401_04.

Johns, M. W. (1991). A new method for measuring daytime sleepiness: the Epworth sleepiness scale. *Sleep, 14*(6), 540-545. https://doi.org/10.1093/sleep/14.6.540.

Johns, M. W. (1992). Reliability and factor analysis of the Epworth Sleepiness Scale. *Sleep, 15*(4), 376-381. https://doi.org/10.1093/sleep/15.4.376.

Johnson, E. O., Breslau, N., Roth, T., Roehrs, T., & Rosenthal, L. (1999). Psychometric evaluation of daytime sleepiness and nocturnal sleep onset scales in a representative community sample. *Biological Psychiatry, 45*(6), 764-770. https://doi.org/10.1016/S0006-3223(98)00111-5.

Kaida, K., Takahashi, M., Åkerstedt, T., Nakata, A., Otsuka, Y., Haratani, T., & Fukasawa, K. (2006). Validation of the Karolinska sleepiness scale against performance and EEG variables. *Clinical Neurophysiology, 117*(7), 1574-1581. https://doi.org/10.1016/j.clinph.2006.03.011.

Kaida, K., Åkerstedt, T., Kecklund, G., Nilsson, J. P., & Axelsson, J. (2007). Use of subjective and physiological indicators of sleepiness to predict performance during a vigilance task. *Industrial Health, 45*(4), 520-526. https://doi.org/10.2486/indhealth.45.520.

Karrer, K., Vöhringer-Kuhnt, T., Baumgarten, T., & Briest, S. (2004). The role of individual differences in driver fatigue prediction. *Proceedings of the 3rd International Conference on Traffic and Transport Psychology*. Nottingham, UK.

Karrer, K., Briest, S., Vöhringer-Kuhnt, T., Baumgarten, T., & Schleicher, R. (2005). Driving Without Awareness. In G. Underwood (Ed.), *Traffic & Transport Psychology – Theory and Application* (pp. 455-469). Oxford: Elsevier.

Kato, M., Sugenoya, J., Matsumoto, T., Nishiyama, T., Nishimura, N., Inukai, Y., Okagawa, T., & Yonezawa, H. (2001). The effects of facial fanning on thermal comfort sensation during hyperthermia. *Pflügers Archiv European Journal of Physiology, 443*(2), 175-179. https://doi.org/10.1007/s004240100681.

Kawahara, J., Sano, H., Fukuzaki, H., Saito, K., & Hiroucht, H. (1989). Acute effects of exposure to cold on blood pressure, platelet function and sympathetic nervous activity in humans. *American Journal of Hypertension, 2*(9), 724-726. https://doi.org/10.1093/ajh/2.9.724.

Khushaba, R. N., Kodagoda, S., Lal, S., & Dissanayake, G. (2011). Driver drowsiness classification using fuzzy wavelet-packet-based feature-extraction algorithm. *IEEE Transactions on Biomedical Engineering, 58*(1), 121-131. https://doi.org/10.1109/TBME.2010.2077291.

Kim, C. J., Park, S., Won, M. J., Whang, M., & Lee, E. C. (2013). Autonomic nervous system responses can reveal visual fatigue induced by 3D displays. *Sensors, 13*(10), 13054-13062. https://doi.org/10.3390/s131013054.

Knijnenburg, B. P., Willemsen, M. C., Gantner, Z., Soncu, H., & Newell, C. (2012). Explaining the user experience of recommender systems. *User Modeling and User-Adapted Interaction, 22*(4-5), 441-504. https://doi.org/10.1007/s11257-011-9118-4.

Koehn, J., Kollmar, R., Cimpianu, C., Kallmünzer, B., Moeller, S., Schwab, S., & Hilz, M.J. (2012). Head and neck cooling decreases tympanic and skin temperature, but significantly increases blood pressure. *Stroke, 43*(8), 2142-2148. https://doi.org/10.1161/STROKEAHA.112.652248.

König, A., Omlin, X., Zimmerli, L., Sapa, M., Krewer, C., Bolliger, M., Müller, F., & Riener, R. (2011). Psychological state estimation from physiological recordings during robot-assisted gait rehabilitation. *Journal of Rehabilitation Research & Development*, 48(4). https://doi.org/10.1682/JRRD.2010.03.0044.

Körber, M., Cingel, A., Zimmermann, M., & Bengler, K. (2015). Vigilance decrement and passive fatigue caused by monotony in automated driving. *Procedia Manufacturing, 3*, 2403-2409. https://doi.org/10.1016/j.promfg.2015.07.499.

Kräuchi, K., Cajochen, C., & Wirz-Justice, A. (1997). A relationship between heat loss and sleepiness: effects of postural change and melatonin administration. *Journal of Applied Physiology, 83*(1), 134-139.

Krajewski, J., Sommer, D., Trutschel, U., Edwards, D., & Golz, M. (2009). Steering wheel behavior based estimation of fatigue. *Proceedings of the 5th International Driving Symposium on Human Factors in Driver Assessment, Training and Vehicle Design* (pp. 118-124). Big Sky, MT: University of Iowa. https://doi.org/10.17077/drivingassessment.1311.

Kregel, K. C., Seals, D. R., & Callister, R. (1992). Sympathetic nervous system activity during skin cooling in humans: relationship to stimulus intensity and pain sensation. *The Journal of Physiology, 454*(1), 359-371. https://doi.org/10.1113/jphysiol.1992.sp019268.

Kreibig, S. D. (2010). Autonomic nervous system activity in emotion: A review. *Biological Psychology, 84*(3), 394-421. https://doi.org/10.1016/j.biopsycho.2010.03.010.

Kuorinka, I. (1983). Subjective discomfort in a simulated repetitive task. *Ergonomics, 26*(11), 1089-1101. https://doi.org/10.1080/00140138308963442.

Lal, S. K., & Craig, A. (2000). Driver fatigue: Psychophysiological effects. *Proceedings of the 4th International Conference on Fatigue and Transportation*. Fremantle, Australia.

Lal, S. K., & Craig, A. (2001). A critical review of the psychophysiology of driver fatigue. *Biological Psychology, 55*(3), 173-194. https://doi.org/10.1016/S0301-0511(00)00085-5.

Lal, S. K., & Craig, A. (2002). Driver fatigue: electroencephalography and psychological assessment. *Psychophysiology, 39*(3), 313-321. https://doi.org/10.1017/S0048577201393095.

Landström, U., Englund, K., Nordstrom, B., & Stenudd, A. (1999). Laboratory studies on the effects of temperature variations on drowsiness. *Perceptual and Motor Skills, 89*(3f), 1217-1229. https://doi.org/10.2466/pms.1999.89.3f.1217.

Landström, U., Englund, K., Nordstrom, B., & Stenudd, A. (2002). Use of Temperature Variations to Combat Drivers' Drowsiness. *Perceptual and Motor Skills, 95*(2), 497-506. https://doi.org/10.2466/pms.2002.95.2.497.

Larue, G. S., Rakotonirainy, A., & Pettitt, A. N. (2011). Driving performance impairments due to hypovigilance on monotonous roads. *Accident Analysis & Prevention, 43*(6), 2037-2046. https://doi.org/10.1016/j.aap.2011.05.023.

Laugwitz, B., Held, T., & Schrepp, M. (2008). Construction and evaluation of a user experience questionnaire. In A. Holzinger (Ed.), *HCI and Usability for Education and Work. USAB 2008. Lecture Notes in Computer Science, vol. 5298.* (pp. 63-76). Berlin: Springer.

LeBlanc, J., Dulac, S., Cote, J., & Girard, B. (1975). Autonomic nervous system and adaptation to cold in man. *Journal of Applied Physiology, 39*(2), 181-186.

LeBlanc, J., Blais, B., Barabe, B., & Cote, J. (1976). Effects of temperature and wind on facial temperature, heart rate, and sensation. *Journal of Applied Physiology, 40*(2), 127-131.

LeBlanc, J., Cote, J., Dulac, S., & Dulong-Turcot, F. (1978). Effects of age, sex, and physical fitness on responses to local cooling. *Journal of Applied Physiology, 44*(5), 813-817.

Li, G., & Chung, W. Y. (2013). Detection of driver drowsiness using wavelet analysis of heart rate variability and a support vector machine classifier. *Sensors, 13*(12), 16494-16511. https://doi.org/10.3390/s131216494.

Liu, C. C., Hosking, S. G., & Lenné, M. G. (2009). Predicting driver drowsiness using vehicle measures: Recent insights and future challenges. *Journal of Safety Research, 40*(4), 239-245. https://doi.org/10.1016/j.jsr.2009.04.005.

Locascio, J., Khurana, R., He, Y., & Kaye, J. (2016). Utilizing employees as usability participants: exploring when and when not to leverage your coworkers. *Proceedings of the 2016 CHI conference on human factors in computing systems* (pp. 4533-4537). San Jose, CA: ACM. https://doi.org/10.1145/2858036.2858047.

Lossius, K., Eriksen, M., & Walløe, L. (1994). Thermoregulatory fluctuations in heart rate and blood pressure in humans: effect of cooling and parasympathetic blockade. *Journal of the Autonomic Nervous System, 47*(3), 245-254. https://doi.org/10.1016/0165-1838(94)90185-6.

Lowden, A., Åkerstedt, T., & Wibom, R. (2004). Suppression of sleepiness and melatonin by bright light exposure during breaks in night work. *Journal of Sleep Research, 13*(1), 37-43. https://doi.org/10.1046/j.1365-2869.2003.00381.x.

Maclean, A. W., Fekken, G. C., Saskin, P., & Knowles, J. B. (1992). Psychometric evaluation of the Stanford sleepiness scale. *Journal of Sleep Research, 1*(1), 35-39. https://doi.org/10.1111/j.1365-2869.1992.tb00006.x.

Mäkinen, T. M., Palinkas, L. A., Reeves, D. L., Pääkkönen, T., Rintamäki, H., Leppäluoto, J., & Hassi, J. (2006). Effect of repeated exposures to cold on cognitive performance in humans. *Physiology & Behavior, 87*(1), 166-176. https://doi.org/10.1016/j.physbeh.2005.09.015.

Mahachandra, M., & Garnaby, E. D. (2015). The Effectiveness of In-vehicle Peppermint Fragrance to Maintain Car Driver's Alertness. *Procedia Manufacturing, 4,* 471-477.
https://doi.org/10.1016/j.promfg.2015.11.064.

May, J. F., & Baldwin, C. L. (2009). Driver fatigue: The importance of identifying causal factors of fatigue when considering detection and countermeasure technologies. *Transportation Research Part F: Traffic Psychology and Behaviour, 12*(3), 218-224.
https://doi.org/10.1016/j.trf.2008.11.005.

May, J. F. (2011). Driver fatigue. In B. E. Porter (Ed.), *Handbook of Traffic Psychology* (pp. 287-297). London: Academic Press.

Mazhelis, O., Luoma, E., & Warma, H. (2012). Defining an internet-of-things ecosystem. In S. Andreev, S. Balandin, & Y. Koucheryavy (Eds.), *Internet of Things, Smart Spaces, and Next Generation Networking. ruSMART 2012, NEW2AN 2012. Lecture Notes in Computer Science, vol 7469* (pp. 1-14). Berlin: Springer.
https://doi.org/10.1007/978-3-642-32686-8_1.

McBain, W. N. (1970). Arousal, monotony, and accidents in line driving. *Journal of Applied Psychology, 54*(6), 509-519.
https://doi.org/10.1037/h0030144.

McIntyre, D. A. (1980). *Indoor Climate.* London: Applied Science.

Mehler, B., Reimer, B., Coughlin, J., & Dusek, J. (2009). Impact of incremental increases in cognitive workload on physiological arousal and performance in young adult drivers. *Transportation Research Record: Journal of the Transportation Research Board,* (2138), 6-12.
https://doi.org/10.3141/2138-02.

Mehler, B., Reimer, B., & Coughlin, J. F. (2012). Sensitivity of physiological measures for detecting systematic variations in cognitive demand from a working memory task an on-road study across three age groups. *Human Factors, 54*(3), 396-412.
https://doi.org/ 10.1177/0018720812442086.

Merat, N., & Jamson, A. H. (2013). The effect of three low-cost engineering treatments on driver fatigue: A driving simulator study. *Accident Analysis & Prevention, 50*, 8-15. https://doi.org/10.1016/j.aap.2012.09.017.

Mercado, J. E., Rupp, M. A., Chen, J. Y., Barnes, M. J., Barber, D., & Procci, K. (2016). Intelligent agent transparency in human–agent teaming for Multi-UxV management. *Human Factors, 58*(3), 401-415. https://doi.org/10.1177/0018720815621206.

Michail, E., Kokonozi, A., Chouvarda, I., & Maglaveras, N. (2008). EEG and HRV markers of sleepiness and loss of control during car driving. *Proceedings of the 30th Annual International Conference of the IEEE Engineering in Medicine and Biology Society* (pp. 2566-2569). Vancouver, BC: IEEE. https://doi.org/10.1109/IEMBS.2008.4649724.

Mieg, H. P. (2006). *Vigilanzentwicklung unter nCPAP-Therapie beim obstruktiven Schlafapnoesyndrom unter besonderer Berücksichtigung der zirkadianen Rhythmik*. PhD thesis. Freie Universität Berlin.

Monk, T. H., Fookson, J. E., Moline, M. L., & Pollak, C. P. (1985). Diurnal variation in mood and performance in a time-isolated environment. *Chronobiology International, 2*(3), 185-193. https://doi.org/10.3109/07420528509055558.

Motor Trend Group (2010). *Automotive Air Conditioning History*. Retrieved from https://www.automobilemag.com/news/automotive-air-conditoning-history/, Access Date: 11.08.2018.

Müller, V. C., & Bostrom, N. (2016). Future progress in artificial intelligence: A survey of expert opinion. In V. Müller (Ed.), *Fundamental issues of artificial intelligence. Synthese Library (Studies in Epistemology, Logic, Methodology, and Philosophy of Science), vol. 376* (pp. 555-572). Cham: Springer. https://doi.org/10.1007/978-3-319-26485-1_33.

Mulder, L. J. M. (1992). Measurement and analysis methods of heart rate and respiration for use in applied environments. *Biological Psychology, 34*(2), 205-236.
https://doi.org/10.1016/0301-0511(92)90016-N.

Murata, A., Fujii, Y., & Naitoh, K. (2015). Multinomial Logistic Regression Model for Predicting Driver's Drowsiness Using Behavioral Measures. *Procedia Manufacturing, 3*, 2426-2433.
https://doi.org/10.1016/j.promfg.2015.07.502.

Nemanick Jr, R. C., & Munz, D. C. (1994). Measuring the poles of negative and positive mood using the positive affect negative affect schedule and activation deactivation adjective check list. *Psychological Reports, 74*(1), 195-199.
https://doi.org/10.2466/pr0.1994.74.1.195.

Neubauer, C., Matthews, G., & Saxby, D.J. (2012). Driver fatigue and safety: A transactional perspective. In G. Matthews, P. A. Desmond, C. Neubauer, & P. A Hancock (Eds.), *The Handbook of Operator Fatigue* (pp. 365-377). Boca Raton: CRC Press.

Niederl, T. (2007). *Untersuchungen zu kumulativen psychischen und physiologischen Effekten des fliegenden Personals auf der Kurzstrecke.* PhD thesis. Universität Kassel.

Nielsen, J. (1994). Usability engineering. Oxford: Elsevier.

Norrish, M. I. K., & Dwyer, K. L. (2005). Preliminary investigation of the effect of peppermint oil on an objective measure of daytime sleepiness. *International Journal of Psychophysiology, 55*(3), 291-298.
https://doi.org/10.1016/j.ijpsycho.2004.08.004.

Nunes, L., & Recarte, M. A. (2002). Cognitive demands of hands-free-phone conversation while driving. *Transportation Research Part F: Traffic Psychology and Behaviour, 5*(2), 133-144.
https://doi.org/10.1016/S1369-8478(02)00012-8.

Olson, L., Cole, M., & Ambrogetti, A. (1998). Correlations among Epworth Sleepiness Scale scores, multiple sleep latency tests and psychological symptoms. *Journal of Sleep Research, 7*(4), 248-253. https://doi.org/10.1046/j.1365-2869.1998.00123.x.

Oron-Gilad, T., & Shinar, D. (2000). Driver fatigue among military truck drivers. *Transportation Research Part F: Traffic Psychology and Behaviour, 3*(4), 195-209. https://doi.org/10.1016/S1369-8478(01)00004-3.

Parasuraman, R., & Riley, V. (1997). Humans and automation: Use, misuse, disuse, abuse. *Human Factors, 39*(2), 230-253. https://doi.org/10.1518/001872097778543886.

Parasuraman, R., Warm, J.S., See, J.E., (1998). Brain systems of vigilance. In R. Parasuraman (Ed.), *The attentive brain* (pp. 221-256). Cambridge, MA: The MIT Press.

Parsons, K. (2002). *Human thermal environments: the effects of hot, moderate, and cold environments on human health, comfort, and performance.* Boca Raton: CRC Press.

Partala, T., & Surakka, V. (2003). Pupil size variation as an indication of affective processing. *International Journal of Human-Computer Studies, 59*(1), 185-198. https://doi.org/10.1016/S1071-5819(03)00017-X.

Pataki, K., Schulze-Kissing, D., Mahlke, S., & Thüring, M. (2005). Anwendung von Usability-Maßen zur Nutzeneinschätzung von Fahrerassistenzsystemen. In K. Karrer, B. Gauss, B., & Ch. Steffens (Eds.), *Beiträge zur Mensch-Maschine-Systemtechnik aus Forschung und Praxis* (pp. 211-228). Düsseldorf: Symposion.

Patel, M., Lal, S. K. L., Kavanagh, D., & Rossiter, P. (2011). Applying neural network analysis on heart rate variability data to assess driver fatigue. *Expert Systems with Applications, 38*(6), 7235-7242. https://doi.org/10.1016/j.eswa.2010.12.028.

Paton, J. F. R., Boscan, P., Pickering, A. E., & Nalivaiko, E. (2005). The yin and yang of cardiac autonomic control: vago-sympathetic interactions revisited. *Brain Research Reviews, 49*(3), 555-565. https://doi.org/10.1016/j.brainresrev.2005.02.005.

Pattyn, N., Neyt, X., Henderickx, D., & Soetens, E. (2008). Psychophysiological investigation of vigilance decrement: boredom or cognitive fatigue?. *Physiology & Behavior, 93*(1), 369-378. https://doi.org/10.1016/j.physbeh.2007.09.016.

Pfleging, B., Fekety, D. K., Schmidt, A., & Kun, A. L. (2016). A model relating pupil diameter to mental workload and lighting conditions. *Proceedings of the 2016 CHI Conference on Human Factors in Computing Systems* (pp. 5776-5788). San Jose, CA: ACM. https://doi.org/10.1145/2858036.2858117.

Philip, P., Sagaspe, P., Taillard, J., Valtat, C., Moore, N., Åkerstedt, T., Charles, A., & Bioulac, B. (2005). Fatigue, sleepiness, and performance in simulated versus real driving conditions. *Sleep, 28*(12), 1511-1516. https://doi.org/10.1093/sleep/28.12.1511.

Phipps-Nelson, J., Redman, J. R., Schlangen, L. J., & Rajaratnam, S. M. (2009). Blue light exposure reduces objective measures of sleepiness during prolonged nighttime performance testing. *Chronobiology International, 26*(5), 891-912. https://doi.org/10.1080/07420520903044364.

Platten, F. (2012). *Analysis of mental workload and operating behavior in secondary tasks while driving.* PhD thesis. Technische Universität Chemnitz.

Pribram, K. H., & McGuinness, D. (1975). Arousal, activation, and effort in the control of attention. *Psychological Review, 82*(2), 116-149. https://doi.org/10.1037/h0076780.

Putilov, A. A., & Donskaya, O. G. (2013). Construction and validation of the EEG analogues of the Karolinska sleepiness scale based on the Karolinska drowsiness test. *Clinical Neurophysiology, 124*(7), 1346-1352.
https://doi.org/10.1016/j.clinph.2013.01.018.

Pylkkönen, M., Sihvola, M., Hyvärinen, H. K., Puttonen, S., Hublin, C., & Sallinen, M. (2015). Sleepiness, sleep, and use of sleepiness countermeasures in shift-working long-haul truck drivers. *Accident Analysis & Prevention, 80*, 201-210.
https://doi.org/10.1016/j.aap.2015.03.031.

Raudenbush, B., Corley, N., & Eppich, W. (2001). Enhancing athletic performance through the administration of peppermint odor. *Journal of Sport and Exercise Psychology, 23*(2), 156-160.
https://doi.org/10.1123/jsep.23.2.156.

Recarte, M. A., & Nunes, L. M. (2003). Mental workload while driving: effects on visual search, discrimination, and decision making. *Journal of Experimental Psychology: Applied, 9*(2), 119-137.
https://doi.org/10.1037/1076-898X.9.2.119.

Reimer, B., Coughlin, J. F., & Mehler, B. (2009). Development of a driver aware vehicle for monitoring, managing & motivating older operator behavior. *Proceedings of the ITS-America* (pp. 1-9). Washington, DC: ITS.

Reimer, B., & Mehler, B. (2011). The impact of cognitive workload on physiological arousal in young adult drivers: a field study and simulation validation. *Ergonomics, 54*(10), 932-942.
https://doi.org/10.1080/00140139.2011.604431.

Reyner, L. A., & Horne, J. A. (1998). Evaluation of 'in-car' countermeasures to sleepiness: cold air and radio. *Sleep, 21*(1), 46-51.
https://doi.org/10.1093/sleep/21.1.46.

Rosenthal, L., Roehrs, T. A., & Roth, T. (1993). The sleep-wake activity inventory: a self-report measure of daytime sleepiness. *Biological Psychiatry, 34*(11), 810-820.

Royal, D. (2003). *Vol. 1. Findings National Survey of Distracted and Drowsy Driving Attitudes and Behavior: 2002* (No. DOT HS 809 566). Washington, DC: U.S. Department of Transportation.

Russell, J. A. (1980). A circumplex model of affect. *Journal of Personality and Social Psychology, 39*(6), 1161-1178. https://doi.org/10.1037/h0077714.

Russell, J. A., Weiss, A., & Mendelsohn, G. A. (1989). Affect grid: a single-item scale of pleasure and arousal. *Journal of Personality and Social Psychology, 57*(3), 493-502. https://doi.org/10.1037/0022-3514.57.3.493.

SAE On-Road Automated Vehicle Standards Committee. (2014). *Taxonomy and definitions for terms related to on-road motor vehicle automated driving systems* (SAE Standard J3016 201401). Warrendale, PA: SAE International.

Sagberg, F. (1999). Road accidents caused by drivers falling asleep. *Accident Analysis & Prevention, 31*(6), 639-649. https://doi.org/10.1016/S0001-4575(99)00023-8.

Sammito, S., Thielmann, B., Seibt, R., Klussmann, A., Weippert, M., & Bökkelmann, I. (2016). Nutzung der Herzschlagfrequenz und der Herzfrequenzvariabilität in der Arbeitsmedizin und der Arbeitswissenschaft. *ASU Arbeitsmedizin Sozialmedizin Umweltmedizin 2016, 51*, 123–141.

Sandberg, D., Akerstedt, T., Anund, A., Kecklund, G., & Wahde, M. (2011). Detecting driver sleepiness using optimized nonlinear combinations of sleepiness indicators. *IEEE Transactions on Intelligent Transportation Systems, 12*(1), 97-108. https://doi.org/10.1109/TITS.2010.2077281.

Saxby, D. J., Matthews, G., Hitchcock, E. M., & Warm, J. S. (2007). Development of active and passive fatigue manipulations using a driving simulator. *Proceedings of the Human Factors and Ergonomics Society 51st Annual Meeting* (Vol. 51, No. 18, pp. 1237-1241). Sage Publications.
https://doi.org/10.1177/154193120705101839.

Saxby, D. J., Matthews, G., Warm, J. S., Hitchcock, E. M., & Neubauer, C. (2013). Active and passive fatigue in simulated driving: discriminating styles of workload regulation and their safety impacts. *Journal of Experimental Psychology: Applied, 19*(4), 287-300.
https://doi.org/10.1037/a0034386.

Sayed, R., & Eskandarian, A. (2001). Unobtrusive drowsiness detection by neural network learning of driver steering. *Proceedings of the Institution of Mechanical Engineers, Part D: Journal of Automobile Engineering, 215*(9), 969-975.
https://doi.org/10.1243/0954407011528536.

Schmidt, E. A., Schrauf, M., Simon, M., Buchner, A., & Kincses, W. E. (2011). The short-term effect of verbally assessing drivers' state on vigilance indices during monotonous daytime driving. *Transportation Research Part F: Traffic Psychology and Behaviour, 14*(3), 251-260.
https://doi.org/10.1016/j.trf.2011.01.005.

Schmidt, E., Decke, R., & Rasshofer, R. (2016a). Correlation between subjective driver state measures and psychophysiological and vehicular data in simulated driving. *Proceedings of the IEEE Intelligent Vehicles Symposium* (pp. 1380-1385). Gothenburg, Sweden: IEEE.
https://doi.org/10.1109/IVS.2016.7535570.

Schmidt, J., Braunagel, C., Stolzmann, W., & Karrer-Gauß, K. (2016b). Driver drowsiness and behavior detection in prolonged conditionally automated drives. *Proceedings of the IEEE Intelligent Vehicles Symposium* (pp. 400-405). Gothenburg, Sweden: IEEE.
https://doi.org/10.1109/IVS.2016.7535417.

Schmidt, E., Decke, R., Rasshofer, R., & Bullinger, A. C. (2017a). Psycho-physiological responses to short-term cooling during a simulated monotonous driving task. *Applied Ergonomics, 62*, 9-18. https://doi.org/10.1016/j.apergo.2017.01.017.

Schmidt, E., Ochs, J., Decke, R., & Bullinger, A. C. (2017b). Evaluating drivers' states in sleepiness countermeasures experiments using physiological and eye data - hybrid logistic and linear regression model. *Proceedings of the 9th International Driving Symposium on Human Factors in Driver Assessment, Training and Vehicle Design* (pp. 284-290). Manchester Village, VT: University of Iowa. https://doi.org/10.17077/drivingassessment.1648.

Schmidt, E., Dettmann, A., Decke, R., & Bullinger, A. C. (2017c). Cold legs do not matter – investigating the effect of leg cooling to overcome passive fatigue. *Postersession Proceedings of the Human Factors and Ergonomics Society Europe Chapter 2017 Annual Conference.* Rome, Italy: The Human Factors and Ergonomics Society, Europe Chapter.

Schmidt, J. (2018). *Detektion der Reaktionsbereitschaft beim hochautomatisierten Fahren.* PhD thesis. Technische Universität Berlin.

Schmidt, E., & Bullinger, A. C. (2019). Mitigating passive fatigue during monotonous drives with thermal stimuli: Insights into the effect of different stimulation durations. *Accident Analysis & Prevention, 126*, 115-121. https://doi.org/10.1016/j.aap.2017.12.005.

Schmidtke, H. (1965). *Die Ermüdung.* Bern: Hans Huber Verlag.

Schömig, N., Hargutt, V., Neukum, A., Petermann-Stock, I., & Othersen, I. (2015). The interaction between highly automated driving and the development of drowsiness. *Procedia Manufacturing, 3*, 6652-6659. https://doi.org/10.1016/j.promfg.2015.11.005.

Schwarz, J. F., Ingre, M., Fors, C., Anund, A., Kecklund, G., Taillard, J., & Åkerstedt, T. (2012). In-car countermeasures open window and music revisited on the real road: popular but hardly effective against driver sleepiness. *Journal of Sleep Research, 21*(5), 595-599. https://doi.org/10.1111/j.1365-2869.2012.01009.x.

Sendowski, I., Savourey, G., Launay, J. C., Besnard, Y., Cottet-Emard, J. M., Pequignot, J. M., & Bittel, J. (2000). Sympathetic stimulation induced by hand cooling alters cold-induced vasodilatation in humans. *European Journal of Applied Physiology, 81*(4), 303-309. https://doi.org/10.1007/s004210050047.

Shahid, A., Shen, J., & Shapiro, C. M. (2010). Measurements of sleepiness and fatigue. *Journal of Psychosomatic Research, 69*(1), 81-89. https://doi.org/10.1016/j.jpsychores.2010.04.001.

Shapiro, C. M., Flanigan, M., Fleming, J. A., Morehouse, R., Moscovitch, A., Plamondon, J., Reinish, L., & Devins, G. M. (2002). Development of an adjective checklist to measure five FACES of fatigue and sleepiness: data from a national survey of insomniacs. *Journal of Psychosomatic Research, 52*(6), 467-473. https://doi.org/10.1016/S0022-3999(02)00407-5.

Sinha, R., & Swearingen, K. (2002). The role of transparency in recommender systems. *CHI'02 extended abstracts on Human factors in computing systems* (pp. 830-831). Minneapolis, MN: ACM. https://doi.org/10.1145/506443.506619.

Stuart, S., Alcock, L., Godfrey, A., Lord, S., Rochester, L., & Galna, B. (2016). Accuracy and re-test reliability of mobile eye-tracking in Parkinson's disease and older adults. *Medical Engineering & Physics, 38*(3), 308-315. https://doi.org/10.1016/j.medengphy.2015.12.001.

Stuke, P. E. B. (2016). *Vertical interior cooling system for passenger cars: Trials and evaluation of feasibility, thermal comfort and energy efficiency.* PhD thesis. Technische Universität München.

Stutts, J. C., Wilkins, J. W., Osberg, J. S., & Vaughn, B. V. (2003). Driver risk factors for sleep-related crashes. *Accident Analysis & Prevention, 35*(3), 321-331.
https://doi.org/10.1016/S0001-4575(02)00007-6.

Swan, M. (2012). Sensor mania! the internet of things, wearable computing, objective metrics, and the quantified self 2.0. *Journal of Sensor and Actuator Networks, 1*(3), 217-253.
https://doi.org/10.3390/jsan1030217.

Tejero Gimeno, P., Pastor Cerezuela, G., & Choliz Montanes, M. (2006). On the concept and measurement of driver drowsiness, fatigue and inattention: implications for countermeasures. *International Journal of Vehicle Design, 42*(1-2), 67-86.
https://doi.org/10.1504/IJVD.2006.010178.

Tham, K.W., Willem H.C., (2010). Room air temperature affects occupants' physiology, perceptions and mental alertness. *Building and Environment, 45*(1), 40-44.
https://doi.org/10.1016/j.buildenv.2009.04.002.

Thayer, R. E. (1989). *The biopsychology of mood and arousal.* New York, NY: Oxford University Press.

Thiffault, P., & Bergeron, J. (2003). Monotony of road environment and driver fatigue: a simulator study. *Accident Analysis & Prevention, 35*(3), 381-391.
https://doi.org/10.1016/S0001-4575(02)00014-3.

Tobii (2018). *Tobii Pro X2-60 eye tracker.* Retrieved from https://www.tobiipro.com/product-listing/tobii-pro-x2-60/#Specifications, Access Date: 08.07.2018.

Toyota Motor Corporation (2018). *Safety Technology History.* Retrieved from https://www.toyota-global.com/innovation/safety_technology/history/#precrash-safety, Access Date: 07.08.2018.

Tran, Y., Wijesuriya, N., Tarvainen, M., Karjalainen, P., & Craig, A. (2009). The relationship between spectral changes in heart rate variability and fatigue. *Journal of Psychophysiology, 23*(3), 143-151. https://doi.org/10.1027/0269-8803.23.3.143.

Tsutsumi, H., Hoda, Y., Tanabe, S. I., & Arishiro, A. (2007). Effect of Car Cabin Environment on Driver's Comfort and Fatigue. *SAE Technical Paper*, No. 2007-01-0444. https://doi.org/10.4271/2007-01-0444.

Van der Laan, J. D., Heino, A., & De Waard, D. (1997). A simple procedure for the assessment of acceptance of advanced transport telematics. *Transportation Research Part C: Emerging Technologies, 5*(1), 1-10. https://doi.org/10.1016/S0968-090X(96)00025-3.

Van Veen, S., Vink, P., Franz, M., & Wagner, P.-O. (2014). Enhancing the vigilance of car drivers: a review on fatigue caused by the driving task and possible countermeasures. In P. Vink (Ed.), *Advances in Social and Organizational Factors* (pp. 516-525). Delft, Netherlands: AHFE Conference.

Van Veen, S., Orlinskiy, V., Franz, M., & Vink, P. (2015). Investigating car passenger well-being related to a seat imposing continuous posture variation. *Journal of Ergonomics 5*(3), 140. https://doi.org/10.4172/2165-7556.1000140.

Van Veen, S. (2016). *Driver vitalization: investigating sensory stimulation to achieve a positive driving experience.* PhD thesis. TU Delft. https://doi.org/10.4233/uuid:89e83a5d-804d-4563-8cd4-6aebc374f24d.

Verwey, W. B., & Zaidel, D. M. (1999). Preventing drowsiness accidents by an alertness maintenance device. *Accident Analysis & Prevention, 31*(3), 199-211. https://doi.org/10.1016/S0001-4575(98)00062-1.

Verwey, W. B., & Zaidel, D. M. (2000). Predicting drowsiness accidents from personal attributes, eye blinks and ongoing driving behaviour. *Personality and Individual Differences, 28*(1), 123-142. https://doi.org/10.1016/S0191-8869(99)00089-6.

Vetter, F., Alber, S., & Wetzel, S. (2003). CO2-Klimaanlagen auf dem Weg zur Serienreife. *ATZ-Automobiltechnische Zeitschrift, 105*(9), 808-816. https://doi.org/10.1007/BF03223497.

Vicente, J., Laguna, P., Bartra, A., & Bailón, R. (2011). Detection of driver's drowsiness by means of HRV analysis. *Proceedings of Computing in Cardiology, 2011* (pp. 89-92). Hangzhou, China: IEEE.

Vidulich, M. A., & Tsang, P. S. (1987). Absolute magnitude estimation and relative judgement approaches to subjective workload assessment. *Proceedings of the Human Factors and Ergonomics Society 31st Annual Meeting* (Vol. 31, No. 9, pp. 1057-1061). SAGE Publications. https://doi.org/10.1177/154193128703100930.

Volkswagen AG (2018). *Technik auf den Punkt gebracht.* Retrieved from https://www.volkswagen.de/de/technologie/technik-lexikon/k-o.html, Access Date: 07.08.2018.

Volvo Cars (2018). *Driver alert system.* Retrieved from https://support.volvocars.com/in/cars/pages/owners-manual.aspx?mc=Y555&my=2017&sw=16w17&article=40b14ac402a747e6c0a801e800bda620, Access Date: 07.08.2018.

Wang, Z., Zheng, R., Kaizuka, T., Shimono, K., & Nakano, K. (2017). The effect of a haptic guidance steering system on fatigue-related driver behavior. *IEEE Transactions on Human-Machine Systems, 47*(5), 741-748. https://doi.org/10.1109/THMS.2017.2693230.

Watson, D., & Tellegen, A. (1985). Toward a consensual structure of mood. *Psychological Bulletin, 98*(2), 219-235. https://doi.org/10.1037/0033-2909.98.2.219.

Watson, D., Clark, L. A., & Tellegen, A. (1988). Development and validation of brief measures of positive and negative affect: the PANAS scales. *Journal of Personality and Social Psychology, 54*(6), 1063-1070.
https://doi.org/10.1037/0022-3514.54.6.1063.

Weinbeer, V., Bill, J.-S., Baur, C., & Bengler, K. (2017). Automated driving: Subjective assessment of different strategies to manage drowsiness. In D. de Waard, F. di Nocera, D. Coelho, J. Edworthy, K. Brookhuis, F. Ferlazzo, T. Franke, & A. Toffetti (Eds.), *Varieties of interaction: from user experience to neuroergonomics* (pp. 5-17). Rome, Italy: The Human Factors and Ergonomics Society, Europe Chapter.

Wyon, D. P. (1973). The role of the environment in buildings today: thermal aspects (factors affecting the choice of a suitable room temperature). *Build International, 6*(1), 39-54.

Wyon, D. P., Wyon, I., & Norin, F. (1996). Effects of moderate heat stress on driver vigilance in a moving vehicle. *Ergonomics, 39*(1), 61-75.
https://doi.org/10.1080/00140139608964434.

Yerkes, R. M., & Dodson, J. D. (1908). The relation of strength of stimulus to rapidity of habit-formation. *Journal of Comparative Neurology and Psychology, 18*(5), 459-482.
https://doi.org/10.1002/cne.920180503.

Young, M. S., & Stanton, N. A. (2002). Attention and automation: new perspectives on mental underload and performance. *Theoretical Issues in Ergonomics Science, 3*(2), 178-194.
https://doi.org/10.1080/14639220210123789.

Zhang, L., Helander, M. G., & Drury, C. G. (1996). Identifying factors of comfort and discomfort in sitting. *Human Factors, 38*(3), 377-389.
https://doi.org/10.1518/001872096778701962.

Zhao, C., Zhao, M., Liu, J., & Zheng, C. (2012). Electroencephalogram and electrocardiograph assessment of mental fatigue in a driving simulator. *Accident Analysis & Prevention, 45*, 83-90.
https://doi.org/10.1016/j.aap.2011.11.019.

Printed in the United States
By Bookmasters